UNDER LION ROCK

UNDER LION ROCK

Jane Houng

QX Publishing Co.

UNDER LION ROCK

Author: Jane Houng
Editor: Betty Wong
Illustrator: Bianca Lesaca
Typesetter: Alan Sargent
Cover Designer: Tina Tu

Published by QX PUBLISHING CO.
8/F, Eastern Central Plaza, 3 Yiu Hing Road,
Shau Kei Wan, Hong Kong
http://www.commercialpress.com.hk

Distributed by The SUP Publishing Logistics (H.K.) Limited
16/F, Tsuen Wan Industrial Centre,
220–248 Texaco Road, Tsuen Wan,
NT, Hong Kong

Printed in Hong Kong by Elegance Printing and Book Binding Co., Ltd
Block A , 4/F, Hoi Bun Industrial Building
6 Wing Yip Street, Kwun Tung
Kowloon, Hong Kong

First edition, July 2023
© 2023 QX PUBLISHING CO.
Text © 2023 Jane Houng
Illustrations © 2023 Bianca Lesaca
Cover photo from *Hong Kong History Excursion: Kowloon Peninsula*,
p.168, by Cheng Po Hung, Commercial Press 2020
'Under the Lion Rock' lyrics © EMI Music Publishing Hong Kong
ISBN 978-962-255-148-0

To my parents
Clara Alexandra Smith (1930–2022)
and Anthony Michael Clarke (1930–2018)

Contents

CHAPTER 1: *A Pocket or Three* 11

CHAPTER 2: *The Cupboard Room* 16

CHAPTER 3: *Make Believe* 21

CHAPTER 4: *A Close Shave* 25

CHAPTER 5: *Magic Wand* 30

CHAPTER 6: *Escape* 36

CHAPTER 7: *When I'm Manly Rich* 41

CHAPTER 8: *A Natural* 47

CHAPTER 9: *The Orange King* 54

CHAPTER 10: *A Discovery* 60

CHAPTER 11: *Sassoon School of Music* 65

CHAPTER 12: *Learning the Ropes and Notes* 70

CHAPTER 13: *Giving Front Teeth* 76

CHAPTER 14: *Sparky Finds a Home* 81

CHAPTER 15: *A Free Weekend* 87

CHAPTER 16: *Charity Begins at Home* 92

CHAPTER 17: *Happy Easter* 98

CHAPTER 18: *Bad Dreams* 104

CHAPTER 19: *Thieving Again* 109

CHAPTER 20: *Horrible Histories* 115

CHAPTER 21: *Joey the Honest Boy* 121

CHAPTER 22: *Dissonance* 128

CHAPTER 23: *Touching Fame* 133

CHAPTER 24: *Dreams and Queens* 140

CHAPTER 25: *The Brag* 144

CHAPTER 26: *Music Is in Your Blood* 150

CHAPTER 27: *Flat as a Pipa Duck* 159

CHAPTER 28: *Under Arrest* 163

CHAPTER 29: *Pre-Performance Antics* 170

CHAPTER 30: *Airborne* 175

Historical Note 181

Music in the soul can be heard by the universe.
　　　　　　—Lao Tzu

I believe that art is the doorway to the divine.
　　　　—*The Pen and the Bell*, Glenn Kurtz

CHAPTER I

A Pocket or Three

THAT EVENING, after school, Joey picked pockets with Todd and Shrimp. They met under the *ding ding* of multi-coloured lights strung high along the embankment of Macau ferry's night market. With a fresh breeze blowing from the South China Sea, it was cool for October. Business at the makeshift stalls was brisk, with buyers sifting trinket-laden trays, sellers spewing prices, and coolies in conical hats carrying baskets on shoulder poles crying *'Ze ze'*.

Joey had already picked a few pockets. He knew it was wrong to steal but what else could he do if his stepfather gave him so little pocket money? Besides, it was an opportunity to meet up with his old school friends. Fingering the coins in his blazer pocket, he calculated how many fish balls he could buy. Ten cents a skewer, five balls a stick. That meant at least twenty.

There, at a candy stall on the edge of the market, was another target. A white boy with orange hair, about the same height as Joey but double the girth. Balancing an interesting-looking object between his knees, he was biting into an explosion of pink candy floss. Joey hummed a snatch of his victory song in anticipation of an easy win.

But Todd and Shrimp had noticed him too. Todd strode ahead, Shrimp trailing him. Joey quickened his pace.

'Watch where you're going,' shouted a swerving cyclist.

Squawk squawk squawked the chicken strung upside down from the handlebars.

The schoolboy, alerted, shoved the case under his arm and made a break for it. But Todd caught up with him, pinning him to the back of a stall.

'Okay, okay,' said the boy. His voice had broken. 'What do you want?' Beads of sweat shone on his forehead.

Joey held him by the collar while Todd rummaged through his blazer.

'My driver is coming,' said the boy, and Joey understood. This rich *gwai zai* had a chauffeur.

'Give me the gun,' said Shrimp in Cantonese, tugging at the case.

The boy shrugged his shoulders, unable to understand.

'I saw him first,' Todd said, elbowing Shrimp out of the way. 'It's mine!'

'*Bang, bang.* You wouldn't even know how to shoot one,' said Shrimp.

'That's no gun,' said the boy.

The case was black, fibreglass, tapered at one end. Why would the boy be carrying a gun? Joey guessed it was a musical instrument. *Click. Click.* Todd flicked the two catches, opened the case, and sniffed.

'*Kapow!*' shouted Shrimp, diving to the ground.

Underneath a yellow cloth was a golden instrument with a flared end. A label inside the case read 'Sassoon School of Music'.

'Duh,' said the boy. 'I told you it wasn't.'

Joey pushed Todd's arm away and lifted the instrument. He'd seen military men blowing one when the British governor – dressed in white from top to toe and wearing a silly feathered hat – saluted the Hong Kong flag. He pressed the pistons. They were cool and smooth under his fingertips.

'What is it?' said Shrimp, leering over Joey's shoulder.

Joey put the mouthpiece to his lips and blew. *Toot toot!*

'That's got my spit on it,' said the boy.

Joey put the instrument back in its case and closed it. He'd always wanted to learn how to play an instrument, especially a loud one, one that soldiers blew, one that could make people jump. Should he take it? Maybe not. The boy – glaring at him now – could be the son of a rich and powerful judge. Anyway, how could he explain it to his parents? Maybe he could flog it to the pawnbroker. No. Even if the instrument was pure gold and worth millions, it wasn't worth the risk. Besides, it was clear that Todd wanted it too.

Honk, honk. Car lights flashed and a Jaguar approached from the side road. 'My driver,' shouted the boy, waving like someone drowning. Todd snatched the case and ran off with it. Joey and

Shrimp ran back into the market, back to their original meeting point under the banyan tree.

Shrimp's skinny legs *kung fu*–kicked into thin air. 'He was small fry,' he said.

Joey pulled Shrimp towards a hawker selling snacks. She was draining a sieve of fish balls into a battered tureen. The water inside bubbled like a boiling river.

'Four skewers please, with extra sauce,' said Joey, salivating.

Wah, were those fish balls hot! His tongue stung with chili and spice.

Shrimp crumpled his brown paper bag, three skewers uneaten. 'I'd better get going,' he said. 'Need to feed my little sisters.'

Joey nodded, distracted. He regretted not pinching the instrument. It had captured his imagination. Only that day, after choir practice at St Thomas's, Mr Lo, his music teacher, had asked him to sing then clap back the rhythms of melodies he'd played on the piano. 'You have an excellent ear,' he'd said. Well, Joey had always rather liked the look of his ears so that was no big deal. But then Mr Lo had told him that learning the violin would be an excellent way to learn about music and that he himself could start teaching him right away. Joey had told him he'd prefer to learn how to sing.

Joey walked towards the bus stop, passing the smelly bean-curd man, the one-eyed fortune-teller and the bicycle repair shop. That night he had a pair of cufflinks to pawn. Made of gold, with a bit of luck. He stopped outside the grilled door, pressed the buzzer and waited for the pawnbroker to buzz him in. The door bolt clicked open and Joey entered a treasure trove glowing red with a family shrine. There was tons of booty inside: wristwatches, ticking clocks, jade bangles, sparkling bejewelled brooches.

The pawnbroker, who had the wispiest eyebrows Joey had ever seen, was serving a non-Chinese guy. There was another man behind the counter clearing up after a recently finished game of Chinese chess. He looked vaguely familiar.

'Wait in line,' said the pawnbroker gruffly, peering through his loupe at a chunky gold necklace. The pendant was square with the Chinese character for water engraved on it. The pawnbroker turned it over and pointed. The seller shrugged. 'China *cheng yu,*' explained the pawnbroker in broken English. 'Ten dollar. Next!'

The pawnbroker weighed Joey's cufflinks on a rusting scale. 'Four dollars,' he wheezed. His tobacco breath smelt as stale as Stepfather's.

'Is that all?' said Joey.

The pawnbroker slapped four coins on the counter. 'Get a job,' he sneered.

The man with the chessboard lifted the lid off his tea cup, swilled his mouth out and spat into a spittoon.

A clock chimed.

'But. . . .'

'No buts,' said the pawnbroker, looking away.

Joey grudgingly took the money. In his heart, he knew the disdain of the leery old geezer was justified. The truth was, he didn't enjoy thieving anymore. And Todd had become such a bully. Joey would be better off finding some new friends. He ran towards the line of buses waiting at the traffic lights. The first one was his. There wouldn't be another one for fifteen minutes and he probably could leap onboard before the lights changed.

CHAPTER 2

The Cupboard Room

J OEY ALIGHTED at the stop a few yards away from the entrance
to his home in Happy Valley. 'Chestnuts, baked chestnuts!'
hollered an old hawker. Joey was hungry but his mama
would be cooking. He pressed the buzzer to open the security
gate of the six-storey apartment block and bounded up the
concrete stairs two at a time towards the second floor. As usual,
Mama had left the front door open to catch some air. The smell
of ginger and garlic sizzling in the wok made his mouth water.

To the rhythm of his 'I'm Home' tune, he unlocked the double deadlock of the lattice gate and slid it open.

'Change your clothes and start doing your homework,' Mama called from the kitchen.

Joey hooked his schoolbag on the back of his bedroom door, swatted an annoying mosquito and switched on the radio for some music. What did he have to study tonight? Chinese, Maths and English. He'd do English first.

Rattle rattle. It was the iron lattice gate. Who was that? Joey went to investigate. A small wiry man was shaking a rod. 'I want to speak to your parents!' he shouted. Joey's heart sank as he recalled where he'd last seen him. But how did he know to come here?

Mama approached, grabbing Joey. 'What's happening?' she cried. 'Ah, Mr Chung!'

Mr Chung wagged a finger. 'I've seen him there for the second time.'

'Where?' said Mama. Joey tried to wrap his arms around her waist but she resisted. All eyes were now on him. 'Joey? What have you been doing?' she said sternly.

'Thieving, I guess. That's what,' said the man.

Mama's dark doe eyes were searching Joey's. 'He'd never do that. Would you, Joey?'

Joey swallowed awkwardly and shrugged his shoulders. But – oh no! – there were footsteps coming up the stairway. Joey immediately knew whose they were. Bitter bile rose from his stomach. The neighbour turned around to see who it was.

'Yes?' growled Stepfather.

'It's Mr Chung from 4C,' said Mama.

'I'm here about your son,' said Mr Chung.

Stepfather's eyes narrowed and his rubbery lips curled downwards. 'My son?'

'Yes! He's pawning things in Sheung Wan!' said Mr Chung.

Stepfather's eyes bulged in their sockets as he struggled for words. 'My son? Who said he was my son?'

Mama gasped, and the sourness in Joey's throat turned into a thick lump he couldn't swallow.

There was a smell of burning and Mama raced back to the kitchen.

'A golden pendant, no less,' said Mr Chung.

'I don't know him!' said Joey.

'Go into your room, boy,' shouted Stepfather.

Joey's cheeks burned with indignation. *And I'm not a pickpocket, I'm not,* he wanted to shout, but didn't.

'Go on then,' yelled Stepfather, lifting an arm. Joey scurried away like a nullah rat.

'Let's sort this out downstairs,' he heard his stepfather say.

Joey threw himself on his bed. He tried to calm his mind by watching the show of shadows on a blank wall but the traffic noise jangled his nerves and the smell of steamed rice made his tummy rumble. Turtle eggs. He just wanted to eat and go to sleep but there'd definitely be no dinner tonight.

'Joey?' His mama was at the door.

'Yes, Ma?'

'Eat, quickly,' she said, passing him a bowl of congee. The rice porridge slid down his throat, warming his stomach. Her tired face was searching his. 'Is there something you haven't told me?' she asked.

Joey's heart beat like a drum. Where could he begin? He shivered inside at the thought of how many secrets he had. 'No,

Ma,' he said firmly, to reassure her, and resolving at that moment never to meet up with Todd and Shrimp again.

'Stealing is a criminal offence, you know,' she said. 'There's a cell in Central Prison for dishonest boys.'

Was that true? Joey lowered his gaze.

'Joey? Joey, look at me.'

His eyes felt heavy as lead.

'I will be very upset if you are lying to me,' she said, 'Because lying to someone is betraying the other person's trust. Do you understand me?'

'I think so,' muttered Joey sadly.

Clomp clomp. Stepfather was ascending the staircase. Joey gulped down a few more spoonfuls before Mama whipped the bowl away and rushed back towards the kitchen.

'Get out of my way,' said Stepfather, stomping towards Joey's room and ransacking his schoolbag. He found paper clips, candy wrappers, a blunt pencil, a wiggly piece of string, four dollars, and twenty cents. 'You steal things, of course you do, you little crook,' he sneered.

Joey edged towards his mama but Stepfather intercepted him.

'Don't hurt him,' pleaded Mama.

Stepfather cupped the coins in front of her, shaking them in her face. 'You think he's innocent? A good boy? My backside!'

Mama stifled a sob.

'Come with me!' shouted Stepfather, gripping Joey's ear to manoeuvre him towards the kitchen. From beside the fridge, he retrieved his long bamboo stick then pulled Joey towards the cupboard room. Joey grimaced. Tonight wouldn't just be a sharp rap on the wrists with an iron spoon but something far, far worse.

'Oh no, Papa, not in there,' wailed Mama.

'And he'll stay until I say so!' yelled Stepfather, pushing Joey into the darkness, slamming the door shut and banging it with the bamboo.

Joey groped for the light switch, switched it on, but there was no bulb in the ceiling socket. Turtle eggs! Stepfather would eat his dinner then come to beat him. His skin smarted at the thought of the *whack whack whack* of the whipping stick on the bare flesh of his bottom. He flung himself against the door envisioning Stepfather eating the steamed fish, leaving Mama only the skeleton to pick at. She was snivelling now, while Stepfather slurped noodles. Since Stepfather had opened his new factory, she cried almost every day. Joey blocked his ears, scrunched his eyes shut and tried to block out the sounds.

'Stop blubbering,' he heard Stepfather say.

'Let me out!' yelled Joey, banging his fists against the door.

'I said, not until I say so!' shouted Stepfather, and belched. Joey imagined him leaning back on the chair, aiding his digestion by rubbing his stomach clockwise with two hands, then tweaking the one long hair growing from a mole on his chin. He kept one fingernail – his pinkie – extremely long, in order to show others he didn't do manual work. 'When I'm not managing my businesses, I buy and sell gold,' he would boast to friends, which sounded impressive until you saw the dingy little office which he shared with old painted ladies who squawked down phones and scribbled numbers.

Red neon signage flashed eerie patterns on the walls. They called this room the cupboard room because of a tall rosewood cupboard which occupied most of the space. Recently Stepfather had been nagging his mama to clear its contents. She wouldn't tell Joey what they were. Joey emitted a cry of frustration and lunged at the door.

'I'm coming, now,' shouted Stepfather, whacking the stick.

CHAPTER 3

Make Believe

STEPFATHER HAD LOCKED JOEY in the cupboard room, again. At least the light bulb had been replaced. Joey crumpled in a heap on the floor, defeated. He'd returned from morning school with the violin Mr Lo had given him, only to be smacked on the palm of his right hand with a ruler. 'I can tell you're not sorry enough,' Stepfather had muttered. Joey could hear him now bickering with his mama. He clenched his fists to see whether it would hurt to bow his violin. *Ow!*

Why was Stepfather so bad-tempered these days? Joey knew stealing was wrong and felt shame for lying. But Stepfather was always angry about something. The previous week he'd been livid about a water leak in their flat. The living room wall was green with mould and the neighbour below had complained about the *drip drip drip* of a leaking pipe, but Stepfather hadn't got it fixed. He hadn't arranged for the installation of an air conditioner in it either, or the cleaning of its chandelier. Did his plastic bead factory in San Po Kong really make lots of money? And what about his new factory? Stepfather had boasted that assembling metal watches with Japanese parts was profitable.

Joey looked over to the cupboard *Wah!* Its padlock was open, beckoning like a crooked finger. His mama must have been sorting it out. Carefully, quietly, Joey unhooked the padlock and one of the doors creaked open. *Aa-choo!* The strong smell of camphor and mothballs tickled his throat. Inside, neatly hung and folded, was a treasure trove of Chinese opera troupe stuff. There were fine silk brocades, long as dresses, for warriors, scholars, and kings. One particularly ornate gown was richly embroidered with phoenixes and dragons, yellow dragons with one, two, three, four, five claws. That meant it was the gown of an emperor. There were masks too, tens of them – black, red, blue, white, some with straggly beards. Joey couldn't remember which faces were for the good guys and which were for the bad. Standing on tiptoes, he reached for an exquisitely carved vanity box. It was packed with face paints, combs, brushes and hair slides. He fingered a comb made of bone. How cool and smooth it felt. He propped up the top of the box, examining miniature side drawers and a mirror. Gently pulling the loop attached to one of the drawers, he found a stash of necklaces, brooches, earrings and rings.

Joey vaguely remembered being taken to a Chinese opera as a child. Once was enough. The screeching voices of the singers sounded worse than chalk scraped across a blackboard. But in the olden days everybody who was anybody went to Chinese opera, his mama said. If he asked, she'd recount stories of the emperors and warriors, maidens and villains. She often listened to performances on the radio. Joey turned his attention back to the smooth silk gowns. They smelt musty and the embroidery had faded but stroking one of them triggered a distant memory. It was of his mama, in the countryside, stirring chopped carrots and corn in a large wok of bubbling pork bones. Sitting on a rock nearby there was a man singing a folk song at the top of his voice. Mama clapped when he finished. 'Where I come from,' the man said, 'Voices have to reach the homes of sweethearts down the valley.' Where was that memory from? Not Hong Kong. All Joey knew about his earliest life was that he and his mama had escaped from the Mainland when he was around three.

Joey's chest tightened. Why did Stepfather want his mama to dispose of these beautiful things? He'd ask her about this just as soon as he could. Dinnertime, hopefully. For now, he continued exploring. On an upper shelf of the cupboard, rolled up in a faded bedsheet, was a collection of stage weapons. Gripping a halberd, he traced the edge of the blade against his leg. It was disappointingly blunt. Brandishing it, he lowered himself into a warrior pose, stamped a foot and took a swipe at an imaginary combatant. He pretended to be a commander ordering his troops to attack the enemy. Stepfather! As Joey swished the long axe to-and-fro, he silently berated him. *Stepfather, you warty toad, you're sure to dock my pocket money for months but I tell you what: I'm going to steal some more. Much more. I'll steal so much money I don't have to take a dollar from you ever again. And I'm going to save and save until . . . until Mama doesn't have to take your money either.* Feeling rebellious now, he put on the emperor's gown, smeared red paint on his cheeks, rubbed it off, put a mask on instead, a blue one with black eyes and a fierce face. He wrapped some jade necklaces around his neck and clipped a crown to his head. Now he was completely ready for battle. All would be resolved. He would strike his stepfather senseless, steal his gold, raid his safe. Then he would whisk Mama to the countryside, to the village where she had grown up, a place where she smiled, a place where the rice grew tall. They would milk cows, raise pigs, pull warm eggs out of hens' bottoms.

'*Waiieeee!*' Joey's Chinese opera voice swooped like a bird in flight.

'Shut up in there,' Stepfather shouted from the dining room.

'Joey, be quiet!' Mama called. Her voice was shrill.

'*Waiiiieeeeeeeeeeee!*' Joey sang, like the Chinese opera singers she listened to. He would fly Mama north. She would cling to him as he swooped through the air like a Chinese dragon. In the land of his dreams, they would build a house, burn joss at ancestor's graves, eat suckling pig.

'Please, no, Baba!' Mama was pleading for Joey's mercy again. But a dining chair rasped against the floor tiles and the enemy was approaching. Crash! The door swung open, banging against the wall, and there was Stepfather's angry face, spittle dripping from his chin. Mama was behind him, loose hair straggled across her face. Stepfather lunged at Joey with the bamboo stick and a sleeve of the gown ripped. Sequins, hundreds and thousands, rained to the floor. Then Stepfather's stick was *swish swish* swishing and Joey suffered the whipped roundness of a sound beating.

CHAPTER 4

A Close Shave

S TAR FERRY HONG KONG SIDE at dusk. The best time to pick a
pocket, with the gentle rocking of the green and white
ferry, the psychedelic neon of distant Kowloon, and the ear-
splitting ring of a bell as a herd of workers surged to beat the
barrier, as final as an executioner's. Along the esplanade there
were touts and pimps, businessmen in suits, *amahs* and rich
kids. Along the pavement cocky Western sailors pestered pretty
Chinese ladies. *Honk! Honk!* A driver jumped out of his saloon

car, waving a fist at a rickshaw coolie who'd pulled out in front of him.

Joey was thieving with Todd and Shrimp again. After more beatings, and no hint of pocket money, he'd decided never to accept handouts again, even if Stepfather offered them. Because there was always an unwritten condition: if Joey accepted the money, he had to obey. Joey was glad to be rid of this obligation, even if it meant fewer snacks.

Joey unbuttoned his blazer, looking for his next victim. Western tourists were the easiest game. They'd coo at the bustling harbour, *ooh* at the gentle sweep of the nine dragon mountains, oblivious to stealing fingers. But American soldiers were to be avoided, unless they were drunk, although Joey enjoyed answering their questions about radios, tattoos, tailors, and shoe makers.

Joey watched Shrimp dodging from trouser to handbag, his bony bare legs light and fast. Todd was tracking a fancy lady who'd alighted from a taxi. All Joey needed to do was run up and bump into her so she wouldn't feel his friend's prying hands. Joey teamed up with Shrimp instead. They sidled up behind a Western couple, the woman in a floral print dress and the fat guy in slacks, open-necked shirt and a cowboy hat. The man's wallet was peeping from his back pocket and the lady's handbag was unzipped. This opportunity was too good to miss.

'Gotcha!' cried the man, snatching Joey's wrist. The cowboy hat fell to the ground.

Joey wrested himself away and made a run for it. But then a policeman was blowing a whistle, and another one taking up the chase.

'Scram!' Shrimp shouted, running hot on Joey's heels. It seemed Todd hadn't noticed.

They bolted towards Queen's Pier, bumping into pedestrians, weaving between them and their surprised faces. When there was no sign of the policemen, Joey stopped to catch his breath. But policemen have the habit of appearing from nowhere. 'Over there,' he called to Shrimp, and they dashed towards the Memorial Shrine, circled it, then ascended the steps leading to the roof of the City Hall's auditorium. From there they had a good vantage point of the area. Hurray! No more policemen, as far as Joey could see. They seemed to have out-foxed them. But you could never be too sure.

Joey jerked his head in the direction of the concert crowd below and Shrimp nodded. Men in Western suits and shiny leather shoes stepped out of chauffeured cars. Ladies in evening dresses and treasure around their slinky necks alighted from rickshaws. There were younger people too, boys with long hair and flared trousers, girls in miniskirts and platform shoes.

They walked across the roof, descended the narrow spiral staircase opposite and joined the people filing inside the building. Joey felt the rush of cool air as he entered the auditorium. People were milling around, smartly dressed and sipping drinks in long thin glasses. He made for the nearest pillar, hid behind it, and studied a nearby poster. It showed a Chinese guy with a troubled gaze. 'Roman Tam', Joey read, the singer in that night's concert. Roman was wearing a glittering gown under the silhouette of the Lion Rock mountain under a full moon. Joey remembered that *Under Lion Rock* was the name of a popular weekly series which his mama used to watch on Aunty Tam's neighbours' TV at the fruit market. The latest version of the theme song played from electronic shops all over town. It had a soft, lilting melody and a gentle beat. Even Stepfather occasionally hummed it in the shower.

A middle-aged Chinese couple was standing nearby arguing. Joey leaned against the pillar, craning his neck to listen.

'It's from him, isn't it, your so-called ex,' the man said. He was dressed top-to-toe in white.

The woman swung her hips and twirled the pearls of her necklace.

Shrimp imitated her action and Joey sniggered.

The man reddened. 'Isn't it?' he repeated, his lower lip quivering.

The woman looked away, batted her eyelashes. 'I can't quite remember.'

The man threw their concert tickets to the floor and marched toward the exit, the woman tottering on her high heels behind him. 'Jimmy? Jimmy?'

Joey picked up the tickets. *Wah!* They'd cost sixty dollars each.

Shrimp grabbed the tickets from him. 'Let's flog them!'

Joey grabbed them back. 'Wait!' Something made him hesitate. Something more than the policemen who may be prowling outside.

'Barbecued goose for me,' said Shrimp.

A bell rang, an announcement that the auditorium was ready, doors opening, the sound of music, people flowing towards it. Two pretty girls swept past holding hands. Joey curled the tickets between his fingers. It was too late to try to sell the tickets and it seemed such a waste to throw them away. Besides, he'd really like to hear Roman Tam sing. 'How about we go inside and listen?' he said. 'It could be fun.'

Shrimp looked doubtful. He pulled up his shorts, lifted a foot, shined his shoes on his knee socks.

Joey straightened his school tie. 'Follow me,' he said.

At the entrance to the concert hall, he felt the usher's eyes looking him up and down after checking the ticket. Turtle eggs, if she made a fuss they should make a run for it. But Shrimp giggled and Joey felt emboldened. He pretended to see his mama inside and waved. As the usher double-checked the tickets, he waved again and faked a smile. 'Hi Mama!' he called.

CHAPTER 5

Magic Wand

WHAT A BUZZING BEEHIVE of sounds! The concert hall was pulsating with people talking and musicians practising. Where was the music coming from? Joey stepped down the aisle to find out. A thick purple velvet curtain hid the stage. Below it was a snake pit of musicians. A violinist waved at Joey and Joey waved back. There were guitarists, drummers and Chinese instrumentalists too. Joey couldn't wait for the show to start.

The lights dimmed, the crowd hushed, and an usher waved an angry finger at Joey who quickly returned to his seat. *Bang! Bang! Bang!* sounded a drum, louder and louder, like a giant heartbeat. The velvet curtains twitched, jerked and opened, as if by magic, and the empty stage looming over Joey was as black as sesame paste. Then spotlights flooded and blinded, people cheered, and the cheers exploded into a gigantic roar at the sight of Roman Tam, head bowed, shrouded in a shimmering gown. Joey gasped as a cascade of river music poured from an electric piano. Shivers coursed up and down his spine when Roman raised his head to sing. His first song was *My Baby*.

Next song, the mood changed. The bass guitarist thumbed a faster riff, the synthesiser player added some jazzy harmony and a saxophonist butted in. Roman clicked his fingers and a clutch of dancers, feathered like peacocks, circled him, strutting and pecking. Joey tapped his feet to the rhythm.

Many more songs followed, in English, Mandarin and Cantonese. Some fast, some slow, some funky, some jaunty. Joey recognised most of them. Shrimp crept out in the middle of *Love You Forever*. His sisters would be hungry. But Joey stayed, spellbound. His voice joined the crowd's shrieks of recognition as the rhythm slowed and the opening strains of *Under Lion Rock* hit the air. Roman clutched his heart, closed his eyes and began to sing.

Life has its joys
But often has sorrows too
When we all meet under the Lion's Rock
At least our laughter exceeds our sighs.

Roman circled the stage, waving his microphone like a magic wand. It was as if he wished to gather everyone up in the wings of his feathery white outfit and whisk them away. He was casting a spell which enchanted all. Joey joined his neighbours swaying from side to side and crying Roman's name.

> *Life has its challenges*
> *It's now without its worries*
> *In the same boat under Lion's Rock we row together*
> *Putting aside our differences and finding common ground.*

The words of the song struck a pebble in Joey's heart and it rippled like a pond. He was rocking in the same boat with the people around him. Then Roman was walking towards them. *Eiya!* Was he waving directly at them? Joey's heart raced.

> *Putting aside our hearts' conflicts*
> *Together we pursue our dreams*
> *In the same boat we promise to go together*
> *Without doubt or fear.*

The tinkle of the piano, more tinkly than Macau ferry fairy lights, or a pocketful of coins. And the honey-sweet sound of the violins, all drippy and delicious. Joey felt his whole being ebbing and flowing with the bittersweet lyrics.

> *Together to the ends of the earth*
> *Joining hands to conquer the challenges*
> *Together we work hard to create*
> *Our everlasting legend.*

The last song ended with violins hovering as high as the hawks in the harbour.

Everyone shouted, 'Encore! Encore!'

'Encore! Encore!' shouted Joey.

Roman looked pained. 'I love you too,' he called to his adoring crowd. The house lights flicked on, Joey blinked and it was time to leave. He floated towards Star Ferry in a sea of goodwill. His eyes were blinking, his ears ringing, his heart singing. And the city lights fizzed like fireworks.

'Where've you been?' asked Mama, when he arrived home. She had flicked off the radio and her arms were folded tightly across her chest. The pungent smell of medicinal herbs bubbled on the cooker. Stepfather mustn't be home yet.

'Did you hear me, Joey?'

'Aw, Ma.'

But Mama's jaw was set. The dimples on her cheeks that Stepfather pinched were nowhere in sight. 'You're all flushed,' she said.

'Singing. I've been singing,' said Joey.

Mama's face softened.

'In a concert.'

'What?'

'At City Hall.'

'City Hall? That's for classical music,' she said.

Was that why there'd been some Western instrumental players? 'But Roman Tam was singing too,' he said, a moment later regretting it. Because the next question would be about the tickets. He had to think fast. 'There was this couple having an argument,' he said (which was true), 'and the man threw the tickets away and Shrimp picked them up,' (half true), 'and—'

'Roman Tam. Did you say Roman Tam?' His mama's eyes were as bright as pomegranate pips.

'And he was wonderful, Mama,' said Joey, taking a breath to sing.

Putting aside our hearts' conflicts
Together we pursue our dreams.

Mama joined in, clasping her hands over his. She had such a lovely voice. Joey sang on, ad-libbing the words he hadn't picked up yet.

'Your voice is as sweet as sugar cane. Just like your papa's,' she said. 'I love Cantopop almost as much as Chinese opera.'

His papa's? A tingle ran up Joey's spine. Why hadn't he worked it out before? He had, he'd known it all along. It was why his mama loved ancient Chinese stories. It was why Stepfather made barbed comments about singers. About actors and artists too, but especially singers. Chinese opera singers in particular. He would switch radio stations or change television channels whenever a Chinese opera was broadcast, and make nasty comments about pretty boys who wore make-up and had to sing for their supper.

All because Joey's father had been a Chinese opera singer!

Then, with the clarity of a Hollywood blockbuster, Joey understood something his mother had said which hadn't made sense at the time. It was Chinese New Year, the Year of the Dragon, when he and his mother had visited an aunty at the upper village of the mountain flank. While the adults played *mahjong,* Joey had climbed towards the summit, scaled the tip of the lion's head and circled its crown. Inhaling the breath-taking views of Kowloon, the Mainland, Victoria Harbour, he'd hollered, 'Joey Wang,' to the adulation of his imaginary fans. The sound of his own name echoing around the top of Lion Rock Mountain had accompanied him all the way down to the squatter village. Back in the hut, he'd whooped, 'I've found my voice!' Mama had caught the eye of Aunty May and laughed. 'Sound familiar?' she'd said.

Questions and answers collided in Joey's busy mind. He didn't know where to begin. 'What *is* Cantopop exactly?' he began.

His mama's warm hands squeezed his. 'Like what Roman Tam sings. The best of the East and the best of the West. Chinese melodies with Western harmony. *Guzheng* with guitars, *dizi* with violin, *erhu* with piano. Your papa would have loved it.' Her face was fixed in a half-moon smile and she blinked slowly.

Joey tossed and turned in his bed that night, digesting everything his mama had told him. She had stopped cooking the Chinese medicine but its smell still irritated his nose. How he wished he remembered more about his papa. All he knew was that he sung with the Jiangsu Kunqu Opera Theatre and met his mama in Canton when on tour. She had brought Joey to Hong Kong after he died. Everything in the cupboard was his.

'Music is a gift,' his mama had said while tucking him up in bed, 'and you have it. That's why you're progressing so well.'

As if from high on Lion Rock, an idea flooded Joey's head. It was as solid as the Lion's ancient body, hard as the granite slab his papa had sat on to sing that folk song to his mama. Joey would become a singer! A Cantopop star, like Roman Tam.

Then he could earn lots of money.

CHAPTER 6

Escape

J OEY WAS UP WITH THE SUN at six sharp. Six twenty-five sharp, out of the door. He liked mornings the best because he could get out of the flat for the day. At nights he would practise his violin and listen to Cantopop to distract the miserable thoughts that frequently filled his head. The Christmas holidays and Chinese New Year had been challenging but he'd enjoyed singing carols at a Christmas service in a church, and visiting Aunty Tam and Uncle Bo at Chinese New Year. But school holidays were over and Joey was back to his school routine.

He was playing the violin in his bedroom one night when Stepfather came home from gambling at the Happy Valley Racecourse. Joey stopped for a moment to take a peek. By the grunting and slouching on the sofa, he guessed Stepfather had lost more than a few hundred dollars.

'Serve me some soup,' he heard.

'It's your favourite. Pork bone with radish,' Mama replied. Joey resumed the G major arpeggio he'd been in the middle of.

Stepfather was dragging his heavy rosewood chair towards the table. 'Tell your son to stop playing,' he ordered.

'But he's practising,' Mama replied.

Stepfather moaned, 'What a racket.'

Mama came to shut Joey's bedroom door but Joey couldn't resist sliding his fingers up and down the fingerboard, simulating the sound of a dying pig.

'Shut up!' shouted Stepfather.

Joey bowed across the upper strings more forcefully. *'Screeeeeeee!'*

His door swung open. Stepfather's face was purple as a beetroot. He grabbed the sheet of music on Joey's makeshift stand and scrunched it into a ball.

'That's Papa's old music,' cried Mama.

'I'll give you Papa,' yelled Stepfather, grabbing another sheet and ripping it up.

'No, please,' cried Joey. Tears sprang to his eyes as he tried to protect the other sheets that Mama had found in the cupboard.

Mama pulled at Stepfather's jacket. But Stepfather pulled away, seizing the violin from Joey and flinging it to the floor. *Smash!* Splintered wood, twisted strings, scattered pegs, the violin was a wreck. Joey wailed in disbelief. Surely it was beyond repair. He picked up what remained of its ebony fingerboard, let it drop again to the floor, and a surge of hatred welled up inside him. 'Get out of my room!' he yelled. 'I hate you, you stupid pig!' The words came pouring out. He'd never said anything so rude before.

Stepfather staggered sideways. Gasping heavily, stinking of *mao tai,* he leaned against the wall to catch his breath.

Mama froze.

Stepfather regained his balance, the corners of his lips bubbling with spittle. 'What did you say?' He raised his arm, ready to strike.

Mama dropped on to her knees. 'No, Baba, hit me instead. It's my fault, blame me.'

Stepfather ignored her. 'Lie down, now!' he yelled.

Joey collapsed on his bed. There'd be another beating, for sure. His backside was already crisscrossed with sores and still painful from the night before last when he'd complained about mouldy smell in the living room. The bamboo had split and shredded, cutting him up so badly that even Mama's special potion hadn't yet healed the bloody weals. He looked around his room. How he hated it, hated its bare walls, its neatly tucked bedding, folded as Stepfather directed. He hated the antiseptic smell of the disinfectant Mama swabbed the tiles with, as Stepfather commanded. His heart raged most fiercely at the sight of the broken violin.

Joey heard Mama begging for mercy in the kitchen, imagined her arms spreadeagled across the corner where Stepfather kept his stick. Joey had witnessed the scene so many times before. But Stepfather must have won because the stick was now being hit against something hard.

'Drink some soup first, Baba, while it's still hot,' pleaded Mama.

Stepfather's chair scraped across the tiles, 'And tomorrow I'm going to turn him into the police for thieving,' he said, slurring his words. 'I know he's still at it.'

Slurp. Slurp. The clink of china and the *chink* of chopsticks against porcelain. The hairs on the back of Joey's neck prickled in anticipation of being hurt. How he despised his stepfather's drinking. Why oh why did his papa have to die? He recalled, in Primary Three, when his mama told him she was going to remarry, a rich man, her boss. Shortly afterwards, they'd moved from Aunty Tam's to a cold, unfriendly flat in San Po Kong, then here, in upmarket Happy Valley, which Joey disliked even more.

Joey suddenly knew what he had to do. He collected the stash of coins hidden underneath his mattress, the packets of snacks hidden in his pillowslip. He went to his desk and scribbled a note:

Dear Mama,
Don't worry. I will telephone you.
Your son,
Joey

He'd thought about running away so many times, especially when the bamboo scaffolding had been erected outside the building to repaint its exterior walls. While doing homework, he would watch the workers swinging from pole to pole with their paint-filled pails. It was now or never.

Joey unzipped his backpack, zipped it back up. Should he go to school as usual the next day but not come back? He opened it again, stuffed his food and favourite comics inside, added a sweater and a pair of shoes. Quickly, quietly, he squeezed through his bedroom window and landed on the balcony. The scaffolding looked solid and square. It was dark but the street lights were bright enough to see. Gripping a thick vertical bamboo pole, he legged it over the iron railing, his stomach lurching at the two-floor drop below. He swung the rest of his body over and placed a foot on a horizontal pole which intersected the main frame. Careful not to lose his footing, he edged his way down, the drop yawning beneath him.

His stepfather's head and shoulders appeared from above. 'Idiot!' he cried.

Mama was trying to pull him back.

Down, down, Joey scrambled faster, scraping a shin, grazing an elbow.

'Help, a robber!' cried Old Aunt Kam who lived one floor below.

Down, down, stepping deftly, Joey felt the bamboo flex and bend under his weight. He slithered down the last pole and took one last look above, only spotting his mama's anxious face. For a moment, Joey wanted to be winched by a rope back up to her.

'He's coming after you,' she screamed, her voice shrill and shocking.

That decided it. 'Bye bye Mama,' he shouted, turning and running down the street.

CHAPTER 7

When I'm Manly Rich

J OEY INSTINCTIVELY RAN towards school, but at the traffic lights, instead of turning right, he went left. He ran without looking back, holding the straps of his backpack to stop it banging against his back. *Beep beep!* A taxi driver waggled his finger at Joey. He'd stepped off the pavement without looking.

The street was bustling with activity. Joey passed a doctor peddling birds' nests, a hairdresser picking an old lady's ear, a man renting comics by the hour. It was hard to believe they had been quietly going about their business while a monster had been destroying his violin. How Joey would like to wrap those broken strings around his stepfather's neck.

Stepfather would call the police, Joey was sure of it. They'd be tracking him down in no time. He needed to get as far away as he could, fast. He imagined disappearing into the rice paddies of the New Territories. Maybe not. Who'd feed him? No, he'd go to Lion Rock instead. A drift of woodsmoke would lead him back to the makeshift hut he and his mama called home before they moved to the wholesale fruit market.

Lion Rock people cared for each other. He would befriend some new immigrants. Someone would look after him. The No. 7 bus went from Star Ferry Kowloon side to the squatter village. Joey remembered the terminus, where the dirt road stopped and a pathway between scrubby bushes led up the mountain. *Clink Clink!* He had almost five dollars in his pocket. It should be enough. He joined a queue for the public light bus to Star Ferry Hong Kong side.

On Star Ferry now, crossing the harbour, with the dark expanse of sea ever-widening. The winter monsoon wind felt calming and cooling. Seawater sprayed his face, refreshing him, clearing his mind.

But at Star Ferry Kowloon side bus stop, Joey discovered the last number seven bus had already departed. Turtle eggs. Where should he go instead? He'd heard of cinemas in Mong Kok where you could sleep undisturbed. He'd travel there and watch cartoons all night. Stepfather forbade him to go anywhere near the area because gangsters were king there. If your car was stolen, triad members would find it for you. They corrupted policemen, smuggled electronic goods over the border, helped hawkers sell drugs. 'Rice cakes for sale,' the street sellers would cry, hiding heroin in the shelves of their food trollies.

Joey needed to eat. His stomach felt gnawed by rats' teeth. He walked along Nathan Road, weaving between coolies and tipsy businessmen. Some electrical shops were still open. The colourful neon lights were in full bloom: Rolex. Longines. Dr Oh's Opal Factory. Susie Tong Night Club. Joey hadn't been on Kowloon side so late before. The smell of street food made his tummy rumble. *Mmm!* Steamed dumplings, deep fried tofu, fish balls, skewers of cows' intestines. He stopped in front of a hawker and her steaming tureen of pork dumplings, jiggling the coins in his pocket. He needed bulk to keep those rats at bay.

'Show her your money,' said a skinny toothless man crouching on the pavement.

'One *siu mai*,' said Joey.

The old man spat down a nullah while the dumplings slipped down Joey's throat, warming it. Still hungry, he ordered some *char siu*.

'What are you doing alone at this time of night?' asked the hawker, a middle-aged woman with bandy legs and a baby sleeping on her back. She handed Joey a free skewer of sizzling green peppers and pork, her dirty sleeve thick with grease and a tight jade bangle circling her wrist. Mama used to wear a jade bangle. Papa had given it to her on their marriage day. In a fit of rage one night, Stepfather had snapped it. Mama had collected all the pieces and hidden them in an old coat.

If Joey still had his violin, he'd play this friendly hawker a tune. He sang her a song instead. She clapped!

A passer-by pointed behind him. 'You've got a fan,' he said.

And there was a small street dog with a cream-coloured coat, sitting on the pavement, head cocked, thumping its tail, as if drawn to Joey's voice. Joey used his teeth to slide a chunk of gristly pork off the skewer and spat it out. The dog pounced.

'Is he yours?' Joey asked the hawker woman.

'No,' she said, 'but he often hangs around here.'

'Does he have a name?'

The woman put her hands on her hips as if to ease her back. 'Well I call him Sparky.'

Joey spat another chunk onto his palm and held his hand out. Sparky's little snout felt dry as a plum.

'You should be going home, boy,' said the woman, stacking her pots and pans. A dark shadow clouded Joey's mind. No, no, no! He would never go back to live with that bully. He'd rather live in a packing crate. He'd find somewhere to sleep around here then travel to Lion Rock at the crack of dawn. All he needed now was somewhere to shelter.

The nearby cinema was only advertising triple-X movies and Joey felt too shy to ask whether there were others. Instead, he crossed the road and walked down a side street. A beggar with a baby was making a bed of cardboard boxes in a shop entrance. Joey kept walking. An off-duty bus man was swaying down the road clutching a bottle of *bai jiu*. Joey stared straight ahead. As the street noise dwindled, he noticed the sound of little paws padding against the tarmac.

It was Sparky, following him!

'Go away!' Joey said, walking faster. 'I said, scram!'

'Woof!'

There were vacant plots of land between buildings now, spaces for construction, demarcated by wire mesh. One plot was being used as a temporary junkyard. Stacks of old vehicles piled high like giant toys. Joey peeked inside. There was a guard slumped on a folding chair under a sheet of tarpaulin fanning his belly with a newspaper. Head nodding, fighting sleep, the man hadn't noticed Sparky squeezing through a rip in the mesh and sniffing a rubbish bag. The rubbish monster Lap Sap Chung needed to come around here and clear up a bit. Joey crouched on his haunches and waited. He'd never been up so late. His eyelids felt heavy from watching static images of the Queen and a fish aquarium showing on two TVs in a nearby *dai pai dong.*

Something jolted him awake. It was Sparky wriggling back, wagging his tail and whining. '*Shhh!*' Joey angrily shooed him away. Undeterred, the little dog trotted down the street, sniffing the pavement. He lifted his leg at a lamppost, and disappeared down an alley.

At last the guard's head stayed down. Joey squeezed through the rip Sparky had used and tiptoed towards a battered truck. The door opened easily. The interior stank but at least he'd found somewhere soft. The leather seats smelt comforting. Joey laid back on them and tried to sleep. A full moon was shining brightly in a cloudless sky. The roar of low-flying planes landing at Kai Tak became less frequent. This bed was more comfortable than one made of packing crates for oranges. That's where he used to sleep when he first came to Hong Kong. Hey! How about going to live there with Aunty Tam, his mother's sister? Her husband was friendly enough. He and his mama had lived in their hut in the wholesale fruit market for around six months. Since living with Stepfather, they only visited her on the third day of Chinese New Year. She was a good cook, making radish cake, and orange sauce for slippery sweet *tang yuan* that melted in your mouth. There was always a wide selection of snacks at Aunty Tam's too – salty crackers, crisps, seaweed, egg rolls. If Joey sang her a song, maybe she would let him choose freely from the crate she kept them in. She didn't have a television but the lady who sold cherries next door did.

Yes, going to live with Aunty Tam was a better idea than Lion Rock. Joey would get well fed. He also could earn some money. He guessed he could work out how to get to Yau Ma Tei.

Eiya! What was that? Something was scrabbling at the door.

It was Sparky. Joey opened the door and the dog flopped on his chest, licked his face and snuggled down beside him, its little heart knocking against his.

The roar of car engines, motorbikes, hooting, shouting. Joey's eyelids felt heavy but he still couldn't settle.

Sparky snuffled. His breath smelt like rotten fish.

'Good night, Mama,' Joey whispered. He sang a little song to her in a quavering voice:

> *Mama, I'll buy you a flat and a car*
> *Mama, I'll buy you clothes and bras*
> *Mama, I'll cook you a silver fish*
> *When I'm manly rich.*

CHAPTER 8

A Natural

WOOF WOOF! Joey woke to Sparky pummelling his stomach. He was in the middle of a dream, raging against Stepfather and that broken violin. How could he? How dare he?

Rat-ta-ra-ra-ta-ra. A workman with a paper hat on his head was attacking the tarmac with a jackhammer. Stripped to the waist, he stopped for a moment to take a deep drag of the cigarette hanging out of his mouth.

Knock Knock! A grim-looking guard was banging at the truck window, 'Get out of here, nuisance.'

Joey grabbed his backpack, scooped a yapping Sparky under his arm and ran away between the car wrecks.

The street was bustling with morning activity. There were men delivering trays of steaming buns, a cyclist balancing baskets of *choi sum*, and street sweepers clearing drains with stiff rattan brushes. Joey approached a hawker selling a variety of snacks from her cart. Cooking oil slopped from the wok. 'Scallion pancakes,' she called. 'Oily and salty and fresh.'

'Three please,' said Joey. As the pancakes sizzled in the wok, he composed a tune in his head.

> *Oily and salty and fresh.*
> *Oily and salty, the best!*

He sat on the kerb and ate, oil dripping on his lap. Two scallion pancakes later, stuffed, he burped aloud to show his appreciation. Sparky was still looking longingly. Joey threw him the third. Sparky sniffed it, and refused it.

'Don't think he likes onion,' said an amah in baggy black trousers and a white *cheong sam* top. She handed Joey five cents. 'Buy him a bone.'

Sparky rolled on his back and woofed, as if to say thanks. Joey laughed aloud. Look at that! Maybe this dog could make money for him. Joey could charge for strokes, and tummy tickles. The proceeds would pay for more than a bus fare.

Two white girls in school uniform stopped by.

'What a sweet dog,' said the one with braids and red ribbons in her hair.

Sparky wagged his tail.

'Does he bite?' said the other, leaning down to stroke him.

'Only my food,' joked Joey. He hadn't heard his English voice for a while.

Sparky ran in circles chasing his tail. *Eiya,* what a show-off!

'Ten cents for a stroke,' Joey said slyly.

The girl pouted. 'I can stroke a dog anytime. I have a poodle.'

'How about a tummy tickle? Twenty cents only.'

The girls broke into giggles. That pricked Joey's pride. Even Sparky looked offended. Maybe his business wouldn't work after all. And he really should be finding his way to his aunty's. But what could he do about the dog? He guessed Aunty Tam wouldn't like him. She kept cats.

'Shoo!' said Joey, pushing Sparky away. He would try and be rid of him once and for all.

Sparky cocked his head.

'I said, SHOO!' said Joey firmly, kicking out at the creature.

The girl with the braids looked alarmed. 'Isn't he yours?' she said.

Joey shrugged, feeling embarrassed. 'Yes. No. I mean—'

'You don't want him?'

Joey paused for a second, his mind whirring. 'He's for sale.'

The girls were giggling again. 'That reminds me of a nursery rhyme,' said the one with braces on her teeth.

Joey knew many English nursery rhymes from kindergarten. Which one was the girl referring to? Remembering it, he sang:

How much is that doggie in the window?
The one with the waggly tail.
How much is that doggie in the window?
I do hope that doggie's for sale.
I don't want a bunny or a kitty.

I don't want a parrot that talks
I don't want a bowl of little fishes
You can't take a goldfish for walks.

As Joey sang, people started gathering. The more people assembled, the less uneasy he felt. This was fun! Meanwhile, Sparky ran between their feet, sniffing their shoes. When Joey finished, everybody clapped.

'You're a natural,' said the girl with braids, handing him five cents. She skipped down the pavement to catch up with her classmate.

Joey considered finding a phone but decided his mama could wait. Better to reach Aunty Tam's first. By the time he reached the fruit market, most of the higgledy-piggledy stalls along the road were open. Sun rays lit up the tiny window of Aunty Tam and Uncle Bo's corrugated iron hut. '*Meeooooow!*' mewed Mei Mei from its roof. Joey remembered the cat's name because of the story he'd overheard his mama telling Aunty Tam on the day she and Stepfather first met. Shopping for oranges, Stepfather had asked her Mei Mei's name then teased her for having cute dimples. 'The cheek of it,' Mama had said, rolling an orange between palms.

'Why did you smile at him, Mama?' Joey had asked.

Mama had tittered. 'Never you mind,' she replied.

How Joey wished she'd hadn't smiled. If Stepfather hadn't spotted her dimples, he wouldn't have offered her a job at his plastic bead factory, she would still be selling oranges and they wouldn't have got married.

Sparky was still sniffing and lifting around a nearby lamppost. Joey stood tall, smoothing his hair flat. 'Aunty Tam?' he called.

Wok and chopsticks in hand, mouth full, Aunty Tam appeared at the window. Her face was as mottled as the fruit she sold, sometimes making Joey wonder whether she and his mama really were sisters.

Sparky was barking and scrabbling at the door. It opened with the click of a bolt and there was Uncle Bo. 'Tie it up over there,' he said, passing Joey some twine from a packing crate.

'Your mother couldn't sleep last night,' said Aunty Tam, wagging a chopstick.

Joey felt bad. Poor Mama!

'Goodness you've grown!' said Aunty Tam.

Joey felt blood rushing to his cheeks.

The aroma of shredded beef with black bean sauce and garlic flooded Joey's senses. They must have been eating the previous night's leftovers. He entered the cosy space that Aunty Tam and Uncle Bo called home, where orange crates were their building blocks: for tables, chairs, as well as the beds. A ceiling fan lazily swung from an electrical wire. Another one was drying clothes on a temporary washing line. Joey removed his backpack, sat on an orange crate and drank the carton of Vitasoy Aunty Tam had given him.

'I must call your Mama straight after this,' said Aunty Tam, tipping some water in a pan.

But I want to stay here,' said Joey, standing up to see what she was preparing for him.

'You're almost as tall as me now!' joked Uncle Bo.

Joey smiled politely. His uncle couldn't stand straight. One leg was shorter than another because a shark had bitten him while swimming to Hong Kong. Joey remembered the story well. Aunty Tam had been swimming in the same group with her fiancé, a man called Lam. When the shark approached,

Lam screamed and swam away, leaving Aunty Tam alone and very vulnerable, especially as she was on her monthly. It was Uncle Bo who had kicked and butted the shark on the nose. After that, Aunty Tam only had eyes for him.

'You must be starving,' she said, ladling Joey some breakfast noodles. Mmm, pork-bone soup. He asked whether Sparky could have some too. In between mouthfuls, he recounted his escape.

'Well Mei Mei remembers you,' said Uncle Bo. For indeed, the cat had jumped on Joey's lap.

When Aunty Tam left for a local shophouse to make a call, Joey busied himself by having a shower.

'Your mama wept when I told her,' she said, upon her return, 'and the old man is mad at her for not agreeing to call the police.'

A surge of anger welled up inside Joey but he managed to control it. 'I'll phone her tomorrow,' he replied. 'But Aunty, I really don't want to go back,' he said.

'You'll have to,' she replied. 'Your stepfather is furious.'

'Then I'm definitely not,' said Joey.

Aunty Tam was picking her teeth with a toothpick. It was now or never. Joey adopted a sweet as Sunkist voice. 'Can I please stay with you for a few days?' he wheedled. 'I could work for you. I'm stronger now.' He flexed his arm muscles to prove it.

'But what about school?'

'I can still go, and work here at weekends.'

Uncle Bo was looking at Aunty Tam. She flicked a spider off her thigh. 'Won't you miss your mama?' she said.

Joey flinched. He wasn't sure, but he didn't feel guilty about running away. Why didn't Mama stand up for herself more? Why didn't she do something to stop Stepfather beating him?

'She guessed I'd be with you, didn't she?' he said. 'She knows I'll be happier here.'

Aunty Tam tutted. 'So you think she'll agree?'

'Oh yes, I'm sure,' he said.

Aunty Tam twitched her eyebrows and laughed.

Joey didn't usually hug. Hugging is not what Chinese people normally do. But at that moment he wanted to bury his aunty in his arms.

CHAPTER 9

The Orange King

WHEN JOEY WAS SMALL, his mama used to sing an old soldiers' song to rouse him:

The sun is up
Get up little piggy!
I come to rouse piggy
Piggy is still on the bed.

Living at the fruit market, Joey never needed waking up. Because before dawn he'd be woken by the grinding of a crane delivering a mountain of fruit for Aunty Tam and Uncle Bo. The few days had turned into weeks.

The oranges came from all over the world and were of varying quality. Before catching a bus to school, Joey would pick out and dispose of the overripe, squashed, mottled, freckled, mouldy or rotten ones. Meanwhile, Aunty Tam would be dividing the saleable ones into two classes, rolling the dark-peeled, heavy and shiny ones into a box on her right, and the thick-peeled, lightweight or pockmarked into a box to her left. These were the ones that poorer people could afford. Uncle Bo would busy himself assembling the metal trellis of their stall.

Once the tarpaulin sheets completely covered their space, selling could begin and the first customers would already be buying by the time Joey left for school. 'Oranges for sale. Who'll come and buy my beautiful oranges?' Aunty Tam would be calling, 'Four for ten cents, six for five.' And Uncle Bo would slosh through the coins in the plastic bucket that swung from a pole to find change. By the time Joey got back from school, the trellis would have been dismantled and Uncle Bo would be sliding the beads of his abacus to calculate the day's earnings while Aunty Tam washed vegetables.

That's when Joey would let Sparky off his lead and walk him to the nearby *char siu* stall where the butcher always had some unsold gristly pork scraps.

Saturdays and Sundays were even more fun because Joey could work all day. He would help the sales by offering brown paper bags to customers fingering oranges. When fewer needed serving, he would compose songs in his head. Whenever he sang them, people would clap and smile. He'd given up playing the violin because the very thought of it triggered unpleasant memories. Besides, he much preferred singing. Through singing, he'd made many friends, such as the friendly fishmonger, a wispy-bearded beef seller, and a bean-curd maker who sold plump white slabs on a rack at the back of his bicycle.

Joey wanted to stay living at the market forever and over the weeks his mama had come round to the idea. Daily life was more peaceful at Happy Valley now. The only problem was Sparky. Aunty Tam didn't like him because he chased Dong Dong, a three-legged stray tomcat who scrounged for shrimps and clams at the wet market. Aunty Tam loved Dong Dong because she'd saved his life after being run over by a car. Since nursing him better, he only let her touch him.

That day was a Saturday and Joey awoke to rain pounding on the roof and Lion Rock heavily shrouded in clouds. Street sweepers were out in force *swish swish swishing* the flooded drains with their stiff brushes.

Sparky jumped on Joey's duvet and licked his face.

'Keep that slimy tongue to yourself,' he said, jumping out of bed. Because Saturday was the day his mama usually visited. What time she arrived depended on Stepfather's moods. Her excuse to leave home was that she was attending typing classes at a secretarial school.

Joey dressed quickly, filling a bucket to wash his face and hosing down the shared toilet.

The street was noisy with the beeps of buses, the cursing of coolies, the shuffle and scrape of flip-flopped feet. Joey settled down to business. Burrowing through a sticky mountain of unsorted oranges, he sang his orange king song:

Joey the king of Kowloon
I will make a fortune soon
Money for Mama and Sparky and me
We will sail the South China Sea.

Aunty Tam was serving a regular customer. 'Eight cents for you, Mr To,' she said.

Clink, clink, clink. She threw Mr To's coins into the plastic bucket.

Shuffle shuffle. Uncle Bo found the change.

'Cherries on special!' called Aunty Ma from the adjacent stall.

Joey stopped sorting the oranges to stroke the money in his pocket. Aunty Tam had changed his coins into two ten-dollar notes, crinkly as the skin of a suckling pig. Earning by working made his chest swell. It felt much better than pickpocketing.

'Woof, woof,' barked Sparky. He was tethered to his lamppost, as usual, but obviously fancied a walk. Joey first drank a carton of Vitasoy to quench his thirst.

Sparky pounced when he first saw Dong Dong, straining against his lead and barking incessantly. Dong Dong hissed, swiping Sparky with a claw, as if to say, 'Look at you. Tied up all day. Not free, like me.' Sparky lunged forwards, snapping his lead, and raced towards Dong Dong, yapping, hot on the cat's three paws.

'Stop!' Joey hollered, to no avail. He'd seen a stray dog kill a cat before and had nightmares of Sparky doing that ever since. Dong Dong disappeared under a stall, Sparky too, and Joey thought he'd lost them both. The snarling and yipping alerted him to where they were fighting.

'Sparky,' he cried, to the rolling-animal flurry of fangs and claws, yowls and hisses. Dong Dong – puffed up as a frizzy hairdo – shook himself free, yowled and bolted.

Joey grabbed Sparky and held him tightly. 'Naughty dog!' he said.

Aunty Tam had heard the racket and was frowning. 'It just won't do,' she said.

It was dinnertime by the time Joey's mama arrived. Her face looked pale and thin. Surprisingly, Aunty Tam didn't mention about the fight and for once, Joey didn't mind too much when asked his average score in Chinese dictation, his worst subject.

'Aw, Ma,' he said.

Mama frowned. 'It's important.'

'Seventy-five. Not too bad, right? I mean what was your average score in Primary Six?'

'Cheeky monkey!' said Uncle Bo.

'I'm a horse, actually,' replied Joey. For indeed, that was his Chinese zodiac sign.

Mama sighed.

A heavy round orange sun hung in the sky. The four of them went to Uncle Chen's, a *dai pai dong* restaurant on the next street, only a couple of minutes' walk away, where food was stir-fried in a giant wok and workmen chinked bottles of Tsingtao beer. Joey ate squid, crispy fried noodles, clams, loofah and egg. Delicious! 'Only two,' piped a fisherman from the TV suspended on the opposite wall. Eeeow, it was that advertisement about family planning again. Joey looked away. Fortunately the next one was about a forthcoming concert.

'Samuel Hui!' said Aunty Tam, mouth gaping half-full.

Sam Hui was a Cantopop star, even more famous than Roman Tam.

'Let's go,' said Joey, rattling coins in his trouser pocket. That magical night at the City Hall was still fresh in his memory and he still couldn't get Roman Tam's songs out of his head.

'That'll be the day,' said Aunty Tam, laughing.

'No, we can, really,' said Joey. 'I'm saving for tickets.'

'You and your singing seem to bring us luck,' said Uncle Bo.

Mama's lips turned downwards. 'Luck, but not money,' she murmured.

Joey's breathed a sigh of discontent. His mama was so difficult to please.

Perched on a stool at the entrance, Aunty Chen surveyed the customers with the eyes of a hawk. *Click click click* clicked the beads of her abacus.

'No money, no talk', as the Chinese saying goes. How long could the Orange King reign?

CHAPTER 10

A Discovery

T HERE WAS A FLASH of lightning, a distant crack of thunder
and splashes of rain streamed down from the tarpaulin.
Aunty Tam was napping, her head propped against a crate
while Uncle Bo served customers.

'Have you finished your homework, little friend?' his uncle
called.

Joey had indeed finished the written part of his Home and
Housing project but needed to do some research outdoors.
He'd chosen to feature Lion Rock, the place which still made

his heart sing. He'd planned to catch a bus there, do some sketching, and with Sparky pawing at the lamppost asking for a walk, now was a good time.

'It should blow over, take an umbrella,' said Uncle Bo.

'*Woof woof,*' woofed Sparky, wagging his tail in anticipation.

A bus came almost immediately. *Ding ding,* rang the bell. Sparky jumped on Joey's lap as the bus lurched forward.

'Get that mongrel off the seat,' ordered the conductor.

An ambulance siren wailed as they neared the famous Kwong Wah Hospital. Joey remembered visiting Stepfather there when he'd had heart trouble. Along Nathan Road there was the stamping and thumping of building work. He pitied the coolies slaving under the torrential rain. They had wrapped themselves in plastic to keep dry.

The bus turned into Waterloo Road and they were travelling towards Kowloon Tong, a residential area for the well-to-do. Lion Rock loomed above, high and mighty, its lower flanks patterned by the shanty towns and tin shacks of the squatter village. Progress was slow because traffic lights stopped and started the flow of traffic. Hey, Sparky! Come back! At a red light, the little mongrel spotted another dog, ran down the aisle, and jumped off the back of the bus. Joey dashed to the exit and jumped off.

'Sparky, come here, naughty boy!' he cried, chasing him.

'*Woof woof,*' woofed Sparky, turning a corner and disappearing out of sight.

Joey gradually gained ground. Sparky's little legs slowed and stopped, allowing Joey to catch him. His ribs were pumping like a concertina. Joey was panting too. The other dog loped away.

Where were they now? Hereford Road, according to a street sign. The road ahead was called Cambridge Road. Both were lined on either side by rows of detached houses, all hidden behind tall concrete walls spiked with broken glass and barbed wire.

A gold-plated nameplate attracted Joey's attention. It read, Sassoon School of Music. The name sounded familiar. A well-dressed Chinese woman wearing gloves and a hat, holding hands with a girl, was buzzing the bell at the entrance. The girl looked Joey's age and was carrying a violin case. She reached down to pat Sparky. 'I like dogs,' she said.

'Come on, Betty,' said the woman.

There was a grinding sound as the metal gate opened. Joey glanced inside. The house and well-manicured garden were the biggest he'd ever seen. There must have been over twenty windows. More children were arriving with their parents. Girls with ponytails or bobs tied with coloured twine. Boys in shorts and T-shirts. No uniform. All carrying instruments. A bespectacled Chinese boy stared at Joey with enquiring eyes.

Joey remembered now. Sassoon School of Music was the name inside the trumpet boy's case. Why would a Western boy have a trumpet from here? Joey put Sparky on his leash and followed a gaggle of Chinese children inside. No one seemed bothered as he followed them down the garden path. The entrance hall of the building took his breath away. It had the grandeur and majesty of the Peninsula Hotel, with high ceilings, marble columns, patterned floor mosaics, and the whirring of ceiling fans. Ahead there was a lift shaft. A descending lift shuddered to a stop, its thick cables locked and its wooden door opened. Out came a white-suited lift assistant. 'Maximum of six, please,' he said. Joey moved forward to enter.

In the lift, Sparky wriggled out of Joey's arms. *'Oooh.* Watch my nylons,' said an English-speaking lady, as Sparky sniffed around her ankles.

'Sorry!' said Joey.

The assistant pressed the button for the third floor and the creaky lift slowly climbed.

A girl stroked Sparky. 'He's lovely,' she said.

The lady with the nylons didn't seem to think so. Joey picked Sparky up.

Ding Ding! There was a grating sound as they arrived on the third floor. The assistant opened the slatted door and pressed his hand against a safety bar to prevent the door from closing. 'Please take care of your dog,' he said, as Sparky struggled to be released.

'Yes, sir,' Joey replied, stepping out.

Joey watched as the mother and daughter walked down the corridor, pressed a door bell and were buzzed inside. The sound of music and voices greeted Joey's ears. He felt curious. He tied Sparky to a water pipe. Then, fighting stomach flutters, he stood up, straightened his backpack, flattened his shirt, and smoothed his hair. When the lift disgorged some Chinese children, he followed them, waited to be buzzed in and entered a spacious reception area with high ceilings, elegant furniture and potted plants. He slunk behind a palm tree, like the time he'd sneaked into the Peninsula with Todd and Shrimp. More children entered, all carrying instruments. Parents were sitting in a waiting area. A large coffee table displayed glossy magazines. Behind a counter, there was a receptionist typing. She didn't seem bothered that he was there. That gave Joey the confidence to sit down too. He picked a bilingual leaflet up about the school. *Sassoon School of Music,* he read, *a specialist music school for talented children. This thriving community of musicians was established in 1970 by generous donations from the Sassoon family. Entry to the school is based solely on musical ability or potential, never on background or ability to pay. The common bond of music makes for an inspirational place which transforms the lives of all who are part of it. We welcome teachers, professional players, composers and conductors, school children and other young musicians to come together to make music. Boarding facilities are available.*

A white-haired man and an elegant white woman appeared from an office. From a photo Joey had just perused, he knew the old man was called Maestro Asimov and the fine lady was his daughter. Deep in conversation, she nodding, he tapping a short white stick against his palm, they walked down the hallway towards double doors. A cacophony of musical sounds escaped when Miss Asimov swung them open.

What was going on in there?

CHAPTER 11

Sassoon School of Music

Miss asimov was walking back to the reception. Joey stiffened. She was now only a few steps away. She smelt of lemons and her pleated skirt was bright lucky red. She handed a document to a typist while gazing in his direction. Joey resisted the urge to run.

'Have you come for a music lesson?' she asked. Her voice was polite and friendly.

Joey's mind reeled with embarrassment and confusion. 'Yes, I have, Miss,' he said, finding his English voice, a voice that surprised him in its boldness.

She continued looking at him as if searching for clues as to who he was. 'What instrument do you play?'

The violin, Joey wanted to answer. But that wasn't true. Anymore. He wished he'd accepted Mr Lo's offer of a new one. But he sang, didn't he? 'I am a singer,' he replied.

'Lovely,' said Miss Asimov. 'I should have guessed from your sweet singsong voice. Now, let me see which room Mr Waters has been allocated today.'

Joey felt his heart pounding as she studied a paper pinned to a notice board.

'Yes, that's right, room twenty-three. It's third along the. . . . Oh, wait a minute. Mr Waters has called in to say he's running late.' She checked her watch, gesturing that Joey should sit down again.

'May I wait in there, Miss?' Joey asked, pointing to the double doors at the end of the corridor where the wash of musical sounds had come from. He'd got so far, he couldn't resist it.

Miss Asimov's crescent-moon eyelids fluttered. 'Why, of course.'

It was a large room with a stage, much smaller than City Hall but just as exciting. Maestro was standing on a podium waving that white stick. And children, lots of children, were playing musical instruments together. In the semi-darkness, Joey made his way between the rows of empty seats. He spotted the girl who'd stroked Sparky in the lift. She was sitting three rows back in a group of violinists. Opposite her, children were playing bigger violins, gripping them between their legs. Behind them, giant violins which children had to stand to play.

Behind the string section, children were blowing instruments, wooden ones, silver ones, black ones that needed a lot of puff, and a long brown one that sounded like a cruise ship hooting in harbour fog. But how sweetly the instruments sang together! It seemed the musical sounds were woven together by the waving of Maestro's stick, whipping them into wild, free, amorphous shapes. Joey would like to hide in one of the huge double bass cases lined up along the back of the stage and listen forever.

Crash! A boy at the back of the stage banged two flattened woks together. Beside the cymbal player, another banged drums. *Bang, bang.* The regular, insistent sound triggered a memory of a sunny day on the beach when Joey was a baby. He was sitting on the sand between his mama, and . . . his papa? Joey couldn't be sure. But he was definitely sitting on his mama's lap eating ice-cream because he could feel the sand sprinkled on her skirt and the coldness of an ice-cream on his lips. *Bang, bang.* A man was beating a drum from a boat and it was approaching the shore. A fisherwoman, her head fringed with black curtains, stepped into the waves. She tipped a bucket of wriggling fish to show his mama and she covered her nose at the pungent smell.

An abrupt pause in the music brought Joey back to reality. Maestro was talking to the leader of the violinists and the ones behind her were scribbling in their music. The music continued and Joey noticed, towards the back of the stage, golden instruments that glinted under the lights. There was one that resembled a car engine, big enough to sit on and drive down a street. Next to the car engine there were two other humongous instruments, the length of the kids who were blowing them, then a pair of much short ones, and. . . . Joey gasped. The orange-headed boy from Macau ferry night market was blowing one of them.

Joey jumped to his feet and turned, with a strong urge to run. But at the back of the hall, Miss Asimov was standing next to a tall man and they were pointing in his direction. I can't escape now, he thought.

At that moment, Maestro tapped his baton on the music stand. 'Oboe, at letter B, you're too soft. It's your solo, Mary. I can't hear it. Lift your bell and let the music sing.'

The blonde oboist, having trouble with her reed, blew down her instrument and it squawked.

'Let's hear it again, in your own time,' said the conductor.

May's beautiful melody floated around the auditorium. It sounded as if it was yearning for somewhere to go.

'Very nice my dear, but the phrases could build even more. Like this,' said Maestro, singing the tune. *'La, la, la, la diddy dah dah. La, la, la, la diddy dah dah.'*

Joey understood what the conductor was asking for. The melody was longing for more movement towards a final resolution sometime in the future. Joey didn't mean to sing. He just opened his mouth and sang the phrases. *'La la, la diddy dah dah. La, la, la, la diddy dah dah.'* His voice was strong and clear and true.

Maestro swung round from the pedestal, shading his eyes from the glare of the overhead lights. He tapped his baton for silence. 'Who was that?'

What on earth had Joey just done? He sat down, stunned by his audacity.

'Carla, who *is* that?' repeated Maestro.

Joey hunched his shoulders. Half of him still wanted to run but the other half had frozen.

Maestro was walking toward him. His hair was dishevelled, his nose hooked, his chin unshaven. But he had kind eyes. Like Miss Asimov's. 'Goodness me,' he murmured. 'What a voice!' He called to the back of the auditorium where Miss Asimov and the tall man were standing. 'Is he one of yours?'

The tall man was a Westerner. He squinted in Joey's direction. 'Er, no,' he answered.

Joey squirmed. Lots of eyes were staring at him now. Western eyes, Chinese eyes. Indian eyes and some nationalities he'd never seen before. Some children started playing their instruments. The oboist was practising that melody. Joey nervously sang the phrase again.

'How did you get in here, boy?' Maestro asked.

Joey laughed nervously. He opened his mouth but no sound emerged.

'Carla, dear,' called Maestro, to his daughter. 'Escort this young man out, please. I'd like to know more about him. Mr Waters, do you agree?'

The man raised his hand. 'Yes, Maestro,' he said. 'Let me audition him.'

CHAPTER 12

Learning the Ropes and Notes

BY THE FOLLOWING WEEK, Joey was an enrolled pupil at Sassoon School of Music. He'd been offered a scholarship and a place in the boarding house. Mama said she was relieved, and Stepfather said he didn't care. After a flurry of shopping for clothes and signing of documents, Joey now lived in a dormitory with five other boys. Ray was his best friend so far. He was thirteen too and had already passed Grade 8 violin. He slept on the adjacent bed and they sat together in the canteen.

Joey's old schools had been deadly boring. Ding dong, registration: Yes Miss. *Ding dong,* Chinese dictation: Bad Joey, lazy Joey, write that character one hundred times Joey. *Ding dong,* Playtime: yeah, play football with friends, swop snacks! *Ding dong:* Maths: Ugh. Only English was a yeah.

The timetable at Sassoon was ten times more interesting. There was much less Chinese dictation but Joey still had to study Chinese books. He'd learned many more characters by having more free time to read. Maths was much easier. English, more difficult, but Joey had the opportunity to speak more, play English games, go to a library and read English books. His favourite subject by far was Music, which comprised a third of the timetable. There were instrument lessons, Music Theory, Music History, and Choir. Joey had to sing, practise and play for three hours a day, which he loved! He even liked being a boarder. It was strictly lights out at 8 pm but no-one checked whether he'd had a shower, or a face-wash, or cleaned his teeth.

Joey had his first proper singing lesson within the week. It was mid-March and a cotton tree was flowering in the beautiful garden outside. Mr Waters was American. That day he was wearing a turquoise batik shirt patterned with exotic flowers. Standing stiff and tall in his plain white shirt, Joey felt unusually nervous. But his new black leather lace-ups shone as shiny as Mr Water's piano.

'My English name Joey Kung. I thirteen year old. Like sing song, dog, fish ball,' he said.

Mr Waters stifled a snigger.

Joey didn't like that. He usually prided himself on his English and used to be top of the class. Teachers would marvel at his vocabulary and lack of accent. His friends joked he'd been a *gwai zai* in a previous life. Once a school chaplain had told him it was a gift from God. Joey attributed his gift to an addiction to English comics and Western pop songs, and the poor introduction he'd just delivered to his nerves.

'I like to sing songs,' Joey said, correcting himself.

'Don't worry,' replied Mr Waters, kindly. 'Your English will improve in leaps and bounds here. Just keep using it at every opportunity.'

'Yes, sir,' said Joey, immediately assured.

'And when do you like to sing?' asked Mr Waters.

'With Roman Tam,' Joey replied.

Mr Waters gave him a questioning look.

'And up mountains, when I have money in my pocket, chicken wings, waffles, chocolate ice-cream, my mama's smile, Tom and Jerry. . . .'

Mr Waters raised an open palm. 'Okay, I've got the idea,' he said, 'I like Tom and Jerry too. But if you want to sing as well as Roman Tam, you'll have to learn how to read music. But hey! Reading Western music is as easy as ABC.'

Mr Waters ran his long fingers along a keyboard, singing as he played: '*A-B-C-D-E-F-G*. These are the white notes on the piano,' he explained.

'And the black notes, sir?' asked Joey, leaning over to tap the ones nearest middle C. The sound of them played in sequence sounded like Chinese opera.

'Ah ha, let me teach you *Chopsticks*,' said Mr Waters.

'*Chopsticks?*' Joey snapped two fingers against his thumb to simulate the action of holding a pair.

Chuckling, Mr Waters hit a C sharp and bounced his fingers up the piano playing all the black notes. Then, with his fist, he hit three black notes together, rocking his lower arm to play others.

Joey couldn't suppress his delight. 'Please play it again, sir,' he said. He squeezed beside his teacher on the piano stool, copying the tune one octave below.

'Well done!' said Mr Waters, moving to Joey's left, the bass side of the piano stool, to add a jaunty accompaniment underneath.

'More fast, sir!' said Joey, laughing. What a jolly song. He loved it. The black notes added flavour to the white notes, like chili sauce.

'You're a natural,' said Mr Waters, after they had completed three rounds of *Chopsticks* in quick succession. 'But this is supposed to be a singing lesson so we'd better start employing those vocal cords of yours.'

What did Mr Waters mean? Something complimentary, because he was grinning.

Joey jumped to his feet and sang the most recent song he'd made up.

'Very nice,' said Mr Waters, clapping. 'Now go and learn how to write it down.'

And now here Joey was lining up in the corridor for another Theory lesson. Theory class was where Joey could learn how to write music. He'd like to learn how to compose. In the last lesson, he'd learned about major and minor keys, treble and bass clefs, semibreves and minims, crochets and quavers. There were so many new words and concepts to memorise.

Joey peeped through the window of the classroom door. Mr Downs had laid blank paper on the wooden desks. He waved Joey in from a piano he was playing. He was a young Englishman with a moustache and Beatles haircut which covered his ears. Joey chose a front desk.

'You're doing well, Joey,' said Mr Downs.

'Thank you, sir,' replied Joey happily. Every lesson felt more interesting than the last.

Mr Downs tucked his shirt into his bell-bottom trousers and adjusted his thick-rimmed glasses as everybody piled in. There was a shuffling and scraping as they settled down. In Joey's old school there had been forty students per class. Here there were only ten.

'Let's begin,' said Mr Downs, 'The paper on your desk is called manuscript paper. Write your name on the top of it.'

Joey studied the paper. Each line was made of lines and spaces where notes could joke and play.

Mr Downs raised his arm to write on the blackboard. *'Add. Beg. Cab. Age.* Can you notate these words?' he asked.

Something clicked in Joey's brain and he wrote feverishly. Of course! A-B-C-D-E-F-G. They were the names of the notes of the lines and spaces. With those notes, he could spell words by writing them on the stave of the treble clef. The word *add* was three notes on a space, line, and line, in the treble clef. The word *bee* was on a line, space, and space. He could tell stories on his manuscript paper: Little *bee*, do you eat *cabbage?*

'How about *babe, face, bead?'* Mr Downs' chalk screeched across the board.

Head down, Joey reached the bottom of the page in no time. *Babe*, your *face* has a *bead* on it. After class, Joey would play this story on the piano. He would sing it too. What fun!

Joey had finished the exercise quickly. *Wah*, it seemed he was the first. He practised twisting his pencil between the fingers of his left hand, until Mr Downs wagged a finger. So instead, Joey turned the page and wrote words of his own on the stave. Every note B in the melody for his mama meant beautiful. Every B in the melody about his stepfather signified bully. Joey wrote other notes below them and listened to how they sounded together. Then, with some deft pencil strokes, he turned the head of a crochet into the head of a stick man. Pretending that each line was a washing line where he could hang things, he drew line extensions and hung socks and panties and vests on them.

'Joey?' said Mr Downs, approaching.

Joey covered his paper with his hand. 'Sorry, sir.'

Mr Downs frowned. 'Move your hand,' he said.

Ray tittered.

'Yes, sir. No, sir. I mean. . . .' stuttered Joey.

Mr Downs tapped the top of Joey's hand with a ruler. 'First warning,' he said.

Joey felt his cheeks burning. Turtle eggs. He hoped his classmates wouldn't tease him for this.

Ding dong, went the bell.

'Good Boys Deserve Favours Always. All Cows Eat Grass. Next lesson I'll teach you the notes of the bass clef,' said Mr Downs.

Joey hurriedly handed in his manuscript and scuttled out of the classroom.

CHAPTER 13

Giving Front Teeth

J OEY HAD TO PINCH himself to believe that he was studying at such a great school. Apart from the music lessons, his favourite, he'd already made four new friends whom he really liked. All were boarders. His best was still Ray – a super-talented violinist but a rather anxious, introverted type. Next best was A So, who slept to Joey's left. A So was shy but friendly and played the flute. Then there was Ben, who slept next to A So – a chubby-fingered viola player and champion of Chinese dictation. The fourth was a girl! Maisy – a cello player from the Mainland who helped him with Maths, Joey's worst subject. Maisy wore her hair in plaits, had unusually long arms, and laughed a lot. Much to his surprise, no one had teased him for having a friend who was a girl.

'What's your last lesson before lunch?' she was asking him now, her body bent to bear the weight of the cello on her back. Joey had given up offering to carry it because she always declined.

He checked the timetable he kept in his shirt pocket. He had a piano lesson, with Miss Wu. Last lesson she'd given him a new album and he could already play most pieces from memory.

Odd. Miss Wu wasn't there. The little room with an upright piano, a piano stool and a chair, felt stuffy. Joey switched on the electric fan but it didn't respond. Maybe she'd gone to reception to complain about it. He flicked through the piano album instead. Every page seemed to tell a different story. *Tom Tom the Indian* – a man with a magnificent crown of feathers at a pow wow – Joey pummelled the keys pretending to dance round the campfire with him. *Emperor Butterfly* – Joey's fingers floated up to where bees were drinking the nectar of sunflowers. *Giant's Causeway* – Joey stamped around the piano's lower reaches.

'Let's hear *Froggy's Fish Cakes*, Joey,' said Miss Wu, entering the room. She was a petite lady who wore body-hugging *cheong sams,* silky stockings and high heels. Sitting so close to her made Joey's head spin. Her good looks had become a topic of conversation in the dormitory.

'It's too easy, Miss,' he said.

'Really? Okay, how about *The Twister?*'

Joey turned to the correct page while Miss Wu settled back into her chair beside the piano stool, retrieving a nail clipper from her handbag. *Clip. Clip.* She was cutting her finger nails! Joey would have a laugh with his classmates about that. He felt the urge to capture her attention. Closing the album, he played the piece from memory then again one tone higher.

Miss Wu slipped the nail clipper back in her handbag. 'That's impressive,' she said.

'What next, Miss?'

'*Tricky Tennis Balls?*'

'No problem.'

Tricky Tennis Balls. Monkey Business. Stormy Waves. Joey didn't have to use the album. He could see the pieces in his head. Like a jigsaw, the notes fitted together under his fingers.

'Well, well,' said Miss Wu, looking a bit flustered. 'I'll bring you the third volume next lesson.'

Someone was knocking at the door. The next pupil, Joey supposed. It was compulsory for every student to study piano and their personal timetables were constantly changing to accommodate the availability of instrumental teachers. The face peering through the small window at eye level looked familiar. Joey swallowed uncomfortably. It was the trumpeter. Joey had since found out his name – Billy. Joey couldn't avoid him this time. Billy was three forms ahead of him and a day boy so their paths rarely crossed. Lunch in the canteen had been risky but Billy always sat with a group of brass players. Toilets were potentially dangerous too. Joey would poke his head round the door first to check the urinals. Only once had he had to dash inside a separate cubicle pretending to do a big job. Orchestral rehearsals on Friday afternoons were dodgy too, when Joey would sit at the back of the hall to listen. The trumpeters sat towards the back of the orchestra and whenever Maestro ordered the brass instruments to play alone, Joey would take a good look at Billy. His face was as round as his trumpet flare and when he blew, his cheeks would balloon like the gills of a puffer fish. At the end of the rehearsals, Joey would quickly leave the hall.

The bell rang. Joey thanked Miss Wu, opened the door and barged through it, head down.

'Look who's here,' Billy said, as Joey passed by. It didn't sound menacing but Joey's body was tensed for action. Ignoring him, Billy calmly entered the practice room and closed the door behind him.

It was bound to happen again. The following day, Joey was walking down the main corridor with Ray when he spotted Billy walking towards them. Joey ducked behind Ray, ran between a row of girls and made a dash for the nearest practice room. He grabbed the handle, charged inside and heaved his shoulder against the door. But Billy had already wedged his foot in the door jamb. 'Got you,' he said, using his bulk to push the door open and pin Joey to a wall. Adrenaline surged in Joey's chest and his arms felt like putty.

'Report him,' Joey called to Ray.

But where were Billy's fists? Why was he smirking? Joey lunged forward.

Billy grabbed Joey's right wrist, deflected the other hand. 'Wait a minute, Small Fry. How about we talk first?'

Talk? That stunned Joey. He wasn't used to that. He said, 'It was my friend who took your trumpet, remember?'

Billy released his grip. 'Yes, of course. I got it back anyway. The police found it in a pawn shop.'

Joey wiped a bead of sweat from his forehead. 'What's there to talk about then?'

'Nothing really, except I could flatten you if I wanted,' said Billy, flexing his arm muscles. 'Watch out!' called Ben and A So who were watching from a safe distance.

'Try me!' said Joey, stiff and angry again. Surely he could knock this fat rich kid to the floor and punch him into submission.

Billy gripped Joey's upper arm and twisted it backwards in a friendly way. 'Why should I? You've got a great voice. Tell you something. I'd donate my front teeth to have a talent like yours.'

Was Billy serious? Joey's anger subsided. Cheekily, he said, 'With no front teeth, would you still be able to blow your trumpet?'

Billy laughed. 'Hah! Good point!'

'What's going on over there?' called an adult. It was Mr Waters.

Billy relaxed his meaty arms and said, 'Buy me some Chinese waffles from that street hawker and we're quits.'

'Okay! It's a deal!' said Joey, stepping backwards for safe measure.

'It's alright, sir,' said Ray.

Billy offered his hand. Joey wondered why. 'Come on,' said Billy. 'Let's shake on it.'

'On what?' Joey still didn't understand.

'Our agreement,' said Billy, laughing.

Shaking hands to resolve differences. How cool is that, thought Joey. It felt good.

'The bell's gone,' said Mr Waters.

CHAPTER 14

Sparky Finds a Home

I T WAS ANOTHER COOL but humid Spring morning. Clouds hung heavily in the sky. After breakfast, Joey checked his pigeonhole for mail. Mama's letter of permission had arrived. The weekend after next was an exeat and boarders were expected to go home. But Stepfather had planned a trip to Canton to pay respects at his father's grave and was insisting Mama accompany him. Rather than stay in school, Joey had persuaded his mama to let him spend the weekend at A So's home.

Joey took the letter to Miss Asimov. Her office overlooked the school gates, which were wide open because day students were arriving. She was looking out of the window.

'What *are* we going to do about that dog?' she said.

Joey couldn't believe his eyes. It was Sparky, trotting around the garden as if he was the owner of the establishment. Joey hadn't seen him since the day of his audition. Before going back to Aunty Tam's, he'd abandoned him at Lion Rock Country Park. He didn't feel bad about it because the moment Sparky had seen other wild dogs, he'd strained to join them, biting his leash and begging to be set free.

'Oh dear,' said Miss Asimov, as Sparky lifted a leg to pee in a flowerbed. The little dog ran over to a washing line where rows of white sheets were drying, snatching and tugging at one of them.

Miss Asimov pushed the window open. 'No, no, stop that animal!' she called.

It seemed her voice didn't carry that far. Miss Asimov motioned towards the door and Joey followed her.

The grass was moist with dew. When Sparky saw Joey, he bounded towards him and rolled on his tummy for a tickle.

'Seems like he knows you,' said Miss Asimov, crouching down to give Sparky a pat.

'I like dogs,' Joey answered, stroking Sparky's velvety ears to try and calm him down.

'I've spotted him a few times now,' said Miss Asimov.

Joey flinched.

'*Woof, woof!*' barked Sparky. *Eiya!* His gums were bright red, swollen, bleeding and his ribs were protruding from a matted and smelly coat. Sparky was injured. Or sick!

'It's the RSPCA for him,' said Miss Asimov, standing up.

'Oh please, Miss. No,' said Joey. Royal Society for the Protection of Cruelty to Animals. That's where many of Hong Kong's strays ended up. Joey had heard that if they weren't adopted within a few days, vets would put them to sleep.

As if understanding what Joey was thinking, Sparky jumped on to his lap.

Miss Asimov looked confused. 'He certainly likes you,' she said.

A crowd of classmates had gathered round. Sparky scrabbled at Joey's chest asking to be let down. He cocked his head and pranced around the garden path as if performing on the street again. Joey's mind raced. What could he do? Instinctively, he thought of Mama. Yes, she would take care of him. But what about Stepfather? He'd forbidden any pets, even singing birds and goldfish. Besides, he'd probably beat Sparky. Or starve him. Joey sighed. The only option was Aunty Tam.

'I think I know the owner of this dog,' Joey lied.

Miss Asimov eyed him suspiciously. Joey had seen that look before.

'I'm sure my aunty would take care of him. She loves dogs. And she lives close, not far away, in Yau Ma Tei.'

Miss Asimov gave him a searching look. Did she believe him? 'But—'

'I have two free periods. I could take him there right now.'

Miss Asimov stood up and brushed some grass seed off her dress. 'Don't be late for choir practice,' she said.

'Yes, Miss.' Joey said.

Joey grimaced. How could he invent stories so easily? They just slipped out of his mouth. He'd have to act fast.

Fortunately, Joey only had to wait a couple of minutes for a bus. The conductor pulled the sliding bars of the gate across the entrance to let them in. *Ding ding!* The conductor tugged a cord. Sparky felt warm in Joey's arms. He snuggled his snout in Joey's elbow crook and closed his eyes.

Joey disembarked at the bus stop nearest to the fruit market. Nathan Road was busy with shoppers and deliverymen and Joey weaved between their carts and baskets. He hurriedly passed a watch-seller with a lens strapped to his eye, a Mister Shoeshine, singing along to the Chinese opera blasting out of his radio. Why were there so many Chinese medicine shops? Bottles of sharks' fins, birds' nests, dried plums. Had Joey taken a wrong turning?

No, at last, the ramshackle stalls of the fruit market. Cherries-on-special Aunty Ma waved when she saw him. And there was Aunty Tam, customers crowding, money basket swinging, brown paper bags flapping in the breeze.

'Joey? What's happened?' she said.

'I think Sparky's sick,' he said.

'Ugh oh,' said Aunty Tam, calling Uncle Bo. Joey would have to win them both over. But how?

Sparky lapped water from the drain near the lamppost.

Uncle Bo's smiles vanished when Joey told them both what had happened. 'Business used to be better with you around but I'm not sure about your dog,' he said.

That gave Joey the idea. He started singing *Under Lion Rock*.

> *Life has its joys*
> *But often has sorrows too.*

'Go on with you,' said Aunty Tam.

'Oh please, Aunty Tam?' said Joey, pulling down Sparky's lower lip to reveal his swollen gums.

Uncle Bo clicked his tongue. But Aunty Tam looked sympathetic.

'Never mind,' said Uncle Bo, motioning to comfort her.

'No,' blurted Aunty Tam. 'Don't tell him.'

Tell him what? Silence. So Joey continued.

> *Life has its challenges*
> *It's not without its worries.*

Aunty Tam was crying! Tears were rolling down her cheeks.

'Dong Dong has disappeared. Your Aunty fears the worst,' Uncle Bo explained.

Joey couldn't help feeling more hopeful for Sparky. He quickly composed a brand new song:

> *Give a little doggy a home*
> *Give a little doggy a bone.*

'You are nothing if not trouble,' said Aunty Tam, suppressing a smile. 'Okay. I suppose it's alright. I'll call your ma to tell her.'

Joey punched the air. 'Thanks so much, Aunty. When I'm a singer, like Roman Tam, I'll build you and Mama a house, and a swimming pool.'

'Me too, please,' called Aunty Ma, spitting a cherry pip into the gutter.

Joey laughed. They would all have houses and swimming pools when he was rich and famous. And Sparky would have a kennel coated in pure gold. Joey would serve him pork bones on a jade plate.

Joey and Aunty Tam went to the beef man who used to give Sparky free scraps. He told them he'd since got a dog to guard his ancestral home in the New Territories and it was fed by his mother. 'So I'd throw these away otherwise,' said the man, scraping skin and gristle off his cleaver and throwing them on to the pavement. Sparky crunched them in the side of his mouth that wasn't sore.

'Are there such things as dentists for dogs?' Joey asked.

'Let me take a look,' said the beef man, prising Sparky's mouth open with an iron skewer.

'Ulcers. I guess he needs some vitamin C,' he said.

'Well, that's one thing I do have,' said Aunty Tam.

Sparky licked the pavement clean then whined for some more meat.

'Thank God he's not a Labrador,' said Aunty Tam.

Joey looked at his watch. The bus would have to fly as fast as a Happy Valley racehorse if he was to reach choir practice on time.

'I promise I'll come back soon,' he said, patting Sparky one last time.

On the bus, Joey imagined himself visiting again next school holiday. He wondered if Mama would be angry about what he'd persuaded Aunty Tam to do. I'll pay them back, he thought, to console himself, but recognising, with a sinking heart, that his interest in remaining an orange king had waned.

CHAPTER 15

A Free Weekend

A SO LIVED ON A HOUSEBOAT in Aberdeen! He warned Joey that he'd have to sleep on the floor.

It was already dusk by the time they arrived. They'd had a full day at school, but no homework. Alighting from the bus, A So walked Joey to a shop selling sausages on sticks. He showed the shopkeeper how many coins he had but the old man shook his head.

'*Sigan,*' said A So. 'If we help Grandpa mend fish nets, he
may give me some pocket money and we can come again
tomorrow.'

Aberdeen harbour was crammed with *junks, kaidoes, sampans,*
motor boats and pleasure boats, all jostling for space. A So was
in a boastful mood. His father used to be a sailor. He'd travelled
all around the world. His uncle had a fish farm in the New
Territories where his father would let him use a knife to chop
feeder fish into pieces. At the water's edge, A So jumped onto
a *sampan,* released it from its mooring and lifted a long pole.
Whoosh! He steered the little boat along a narrow waterway
snaking between the creaking beams of hulls. The Jumbo
seafood restaurant, high as a *tong lau,* towered above.

A So seemed to know everyone!

'Hello, Aunty, hello Uncle,' he called as his neighbours folded
washing, peeled vegetables and scaled fish at the back of their
boats. Joey marvelled at the potted plants and washing lines,
buckets and poles, sizzling stoves and chickens in coops. An
aunty held her nose, hawked and spat in the sea – yuck!

They weren't too late to help A So's grandma clear her fish
traps and sell some fish.

'Jump in,' she called from a rocking *teng.* She was barefoot
and wore a conical bamboo hat. *Sputter sputter,* the boat
zigzagged between traffic. Stopping at the side of restaurants
on water, she used a pole to hook the traps. The nets were
teeming with little grey fish which had been feasting on the
leftovers thrown in the sea.

'Oh look, a baby turtle!' shouted A So.

Grandma let it go, only tipping fish into a water tank. Then
they wended their way from boat to boat selling them. *Ding
ding!* Buyers threw the money into a plastic bucket and A So

did the Maths for change. The strong smell of petrol and fish made Joey feel sick but he didn't complain.

Dusk drew shadows and at sunset the smell of dinners wafted in the wind. A So's mother was crouching at the back of their junk, cooking. 'Coming your way, your favourite sweet and sour pork,' she shouted to A So above the roar of the wok. 'Go and fetch your sister.'

Joey had to concentrate on keeping his balance while climbing the wooden ladder to the roof. Little Ying Ying was tied to a rope there which was knotted to the main mast. She jumped with joy at the sight of her big brother.

It was dark by the time all the family had gathered around the table. Strings of electrical lights flickered spookily on the wall-hung photos of A So's ancestors. Joey ate hungrily.

'Hey, Ying Ying can use chopsticks now,' said A So, placing some more *choi sum* in her bowl.

'*M'goy ge ge,*' replied Ying Ying.

'After dinner, you must perform for us,' said A So's grandpa. His face was wrinkled and worn and two of his front teeth were missing.

'Don't wait for me,' called A So's mother, stacking bowls.

A So chose one of his grandpa's *dizis*. He played a jaunty melody which his grandpa had taught him. Then Joey sang a verse of *Flying Dagger*.

A So's father rubbed his stomach and farted. 'Enough, now,' he said, reaching for the *mahjong* case. Family members gathered around the table to play, shuffling the tiles with waves of their arms. *Clickety-clack*, the tiles were stacked as walls in front of them and the game could commence. Joey joined the children in a card game of diving turtle. When Ying Ying fell asleep on A So's lap, his classmate carried her to where a baby was already

fast asleep on the teak floor. Lying beside his two younger
sisters, he patted their backs.

Joey lay beside A So. 'You can share my duvet,' his friend
whispered.

There was some tugging to-and-fro for space, but soon the
five of them were curled up and snuggled close. *Clack clack
clack.* Would Joey ever be able to get to sleep? A So was already
breathing steadily. The last thing Joey remembered was slapping a mosquito.

Next morning, A So and his sisters were nowhere to be seen.
Joey quickly changed into his shorts and T-shirt and went to
find A So. He was on the roof mending fish nets with his
grandpa. It was a bright, sunny day. Grandpa immobilised his
net with a foot while wrapping thick twine around its metal
frame. A So was sitting cross-legged beside him threading
needles. 'Like this,' he said, throwing Joey a ball of string.

A So leaned towards his grandpa and whispered something
down his ear. The old man reached into his trouser pocket and
passed him a couple of coins. A So winked and Joey followed
him down the ladder.

'Come back soon,' called Grandpa.

The sausages were too spicy for Joey's liking but he ate them
anyway. A girl waved from a stall where a man was banging a
metal pot into shape.

'That's Mary, my girlfriend,' boasted A So, waving back. She
had short hair and lanky legs.

Mary crossed the road to join them.

'Do you want to come to my new home?' she asked shyly.
Her family had lived on a boat next to A So but they'd been
resettled into public housing.

'You get used to the stairs,' she said merrily as they climbed yet another flight, 'And I've made lots of new friends.'

The flat was small, square, and very clean. Mary's parents were out working. Her two younger brothers were cutting out pictures from magazines for a school project. Mary retrieved a can of Coca-Cola from the fridge and poured equally into three beakers. 'I don't think Mama will notice one's missing,' she said.

The Coca-Cola was ice cold and delicious. Joey enjoyed the feeling of liquid slipping down his throat, the whir of the electric fan, the view of the bustling harbour. A So had joined the two boys and was snipping out a picture of a posh new tower block. '*Wah,* I'd like to live there,' he said.

Joey shook his head in disagreement. He'd much rather be living surrounded by caring family and friends somewhere like here, or A So's boat, rather than at Stepfather's with its chandeliers, air conditioners and television. Why was he so obsessed with showing off his wealth? Why did he boast that he could buy a Rolls-Royce and eat abalone every night, if he wanted? Nobody talked about getting rich at school. And there was Aunty Tam, never mean-spirited, always jolly, but too poor to buy herself a new coat last Chinese New Year. One thing was certain: money didn't buy you happiness.

CHAPTER 16

Charity Begins at Home

MISS WU PRESENTED JOEY with a fourth album. There were no illustrations but many more dots and dashes, lines and squiggles. Joey noticed a list of odd-looking words listed in a glossary. Many of them ended with the letter 'O'.

'Presto, largo, allegro, adagio.' Joey twisted his mouth into strange shapes.

Miss Wu translated, 'In Italian that means very fast, slow, fast, very slow.'

'Piano, legato, staccato,' Joey continued.

'Quiet, smooth and short.'

'Why the Italian language?

'Italians were the first to use expression marks,' said Miss Wu.

'But why?

Miss Wu crumpled the newspaper she'd used to collect her fingernail clippings. She was making a habit of this and it repelled Joey, despite her curvaceous body.

'They say Italians are very expressive people,' she said, with a sigh.

'Like you?' Joey almost said, but didn't. Miss Wu, the only teacher who wore makeup and splashed perfume on her dresses. Ben had recently shared that he'd had a sexy dream about her.

Joey went back to the glossary. From now on, he'd write expression marks on his compositions too. For the songs that flooded his head.

That night, after supper and homework, Joey went to the games room to challenge A So in a table football match. Unusually, A So wasn't there. Maybe he was practising that passage that Maestro had asked him to play by himself during the orchestral rehearsal that afternoon. A So had fluffed some notes – always so embarrassing. In the common room, Maestro and Miss Asimov were chatting with a group of sixth-formers. Joey joined the group, listening to why the music school was called Sassoon School of Music. It was founded by the Sassoon family who were Jewish and came from Iraq. They moved from their home country in order to escape religious persecution. Joey would study an atlas to find out exactly where Iraq was. Somewhere in the Middle East, he reckoned.

'The Sassoon family used to export silk and tea from China in exchange for opium,' Miss Asimov continued. 'Opium grew in India, where boats docked on their way to Asia. It was a very profitable business.'

Joey knew about opium. He'd smelt the distinctive odour of addicts 'chasing dragons' in an opium den on Lion Rock.

'There's a road in Pok Fu Lam named after them, Sassoon Road.'

Wah. That's how you got a road named after you.

'Are you from Iraq too?' Joey asked.

'Oh no,' said Miss Asimov, dropping her gaze and picking up a stray thread from the worn carpet. She obviously missed her home country, a lot. 'We're from Russia, or the USSR, as it's called now. We are Russian Jews. White Russians.'

Because of skin colour? Miss Asimov's skin was as white and smooth as tofu. She was beautiful, Joey thought, apart from that honker of a nose.

A girl asked the question in Joey's mind. Miss Asimov shook her head. 'White because we were against the leaders of our country. First, the tsar, and then the Bolsheviks.'

'But why did the Sassoon family use their money to set up a music school?' Joey asked. He couldn't imagine his stepfather doing anything like that with all his millions.

'Charity begins at home,' replied Miss Asimov. 'Have you heard of that saying? Also, Joey, because the Sassoon family were very religious. They felt guilty about trading drugs so they donated generously to good causes.'

'David Sassoon recorded his vision of a specialist music school in a journal,' continued Maestro, selecting a book from the bookshelf and turning to a well-leafed page. ' "The school will be free for musically-gifted children regardless of their background. It will transform the cultural desert of Hong Kong into a fertile oasis of creativity," ' he read.

'And seeds have been sprouting ever since,' said Miss Asimov. 'Even the Hong Kong Philharmonic Orchestra went professional four years ago.'

I must go to one of their concerts, thought Joey.

Maestro tipped his mug and finished his cocoa. 'Before we go to bed, let me tell you a funny story,' he said. 'When the Philharmonic Orchestra was still amateur, I was called in to be the leader at a fundraising concert. As you know, the lead violinist sits to the immediate left of the conductor, a man called Lim Kek-tjiang at that time. Well, my getting paid had caused some discontent within the ranks but that was a condition of my service. I'd recently arrived in Hong Kong and needed money for food. Anyway, there I was trying my best to prove my worth, and playing rather well, even though I say it myself. But Maestro Lim's gaze was permanently fixed in my direction as if checking my every note. You can imagine how uncomfortable I felt. I only found out later that his right eye would focus straight ahead but the left pupil was always turned outwards.

Joey collapsed on to the carpet with his classmates, convulsed in laughter.

As usual, Miss Asimov entered the dormitory a few minutes before lights out. Joey quickly scribbled the last few bars of a song called *Homesick* he'd been composing, inspired by Miss Asimov's sad gaze at the threadbare carpet. 'For you, Ray,' he said, on a whim, because anything he wrote for soprano could be played on a violin. Ray read the expression mark. 'Andante cantabile, singing at a walking pace, I like the sound of that,' he said, stroking his instrument. He loved his violin so much he slept with it.

Miss Asimov had overheard their conversation. 'It's so nice of you to compose something for Ray,' she said, tousling Joey's hair. Creepy. Even his mama didn't do that. His teacher smelt as sweet as ginger flowers. Should he tell her that she was the inspiration?

Time for lights out. Before her final round, Miss Asimov switched on RTHK, the classical music station with plummy English announcers. Then the singer was soaring, as if hovering on the wings of hawks circling the Hong Kong sky.

Ray was tossing and turning by Joey's side.

'Can you sleep?' Joey asked.

'I feel mum-sick,' he replied, turning his head away.

Mamas. The last time Joey had seen his was when she and Stepfather had come to sign the scholarship papers. She'd worn a pair of new shoes. High heels. 'Your son will realise his potential here,' Maestro had said grandly, and Stepfather had tweaked his long mole hair.

Mama. Confucius said it was his duty to look after her his whole life. And he would. When he was manly rich. Joey fingered her photo, the one he hid inside his pillow case, the song on the radio carrying him back to her. He was folded under her wings, tucked inside her warm breast.

'Well don't you, miss yours, I mean?' Ray whispered.

Sure he missed, but not a lot, and he dreaded seeing his Stepfather again at term break. 'A little bit,' he replied.

Miss Asimov was standing at the foot of Ray's bed. 'Still talking?' she said gently.

Ray cuddled closer to his violin. 'I want to see my mother,' he said.

'Come to the office first thing tomorrow morning and we'll call her, dear,' said Miss Asimov.

The singer's voice was floating higher and higher. Joey wondered what the words meant. But it didn't matter. Because he and his mama were flying together, flying away from Stepfather, in one of those aeroplanes that flew over the school, rattling the window panes. They were sitting beside each other in an aeroplane, drinking soda. She was wearing those new high heels and fingering a foreign passport in her leather handbag. England, that's where they were going. To start a new life, like Maestro and Miss Asimov had. Mama had told him she wanted to emigrate a few times. It was many a Hong Konger's dream.

CHAPTER 17

Happy Easter

I T WAS THE EASTER HOLIDAY and the weather was cloudy and grey. Joey had slept late hoping his stepfather would have left for work before he woke up. But Stepfather was still home, too sick, or drunk, to go to his office. Joey tiptoed past his half-closed bedroom door. A radio was spouting stock prices. Turtle eggs! The book Joey was carrying had scraped against the hallway.

'Is that you, Maggie? Tea. I need more tea.'

'I'm coming,' called Mama from the kitchen.

Joey froze.

'Joey? It's Joey, isn't it? Come here!'

Stepfather was sitting up in bed reading a newspaper. The room stank of stale tobacco and garlic.

'Come closer, boy,' he said. His face was puffy and red and his rheumy eyes narrowed and hardened as their eyes met.

'Give up this stupid idea of yours,' he said. 'Singing? What utter nonsense. Singers are entertainers, the lowest class of people. The entertainment industry is full of gangsters and immoral women. If I hear any more about your ridiculous ambition, I'll pull you out of that fancy school.'

'But—'

Stepfather hawked and spat in his spittoon. '*Zou daai si,* a man has to make money, do big things in life. He has to make business, own properties, drive foreign cars. No money, no talk.'

'No money, no talk,' Joey repeated, shuddering inside. Did Cantopop stars talk like this?

'That's right,' said Stepfather, banging his fist against the newspaper.

Mama was pouring boiled water into Stepfather's flask on his bedside table. 'Miss Asimov told me his English is outstanding,' she said. 'Maybe he could get a civil service job.'

'Or Jardine's, Wheelock or Swire,' added Joey.

Stepfather scoffed. 'Your son has had it too easy, that's the problem,' he said. 'The year before you came to Hong Kong, the leftists were running riot here. They shot five Hong Kong people dead. There were mass stabbings and shootings. Bombs. How could this new generation defend us if the Mainlanders tried to take power again?'

'I could learn some *kung fu?'* said Joey, with a Bruce Lee-style side-kick.

'Bah!' said Stepfather. 'Teach your son some sense, will you, *lao po?* And come on, admit it, you wouldn't have married me if I wasn't rich, would you?'

Mama looked as if someone had slapped her across the face. 'Don't forget I make my own money too, Baba,' she mumbled, splashing tea in Stepfather's mug.

'Thanks to me,' said Stepfather.

Joey's heart raged. 'When I'm grown up, I *am* going to be a singer,' he said hotly.

Stepfather couldn't control his anger. 'Take him away from me!' he yelled.

Joey slunk back into his room. This holiday was getting worse by the minute. On the bus the previous evening, when he'd told Mama how pleased Mr Waters was with his progress, how he'd been picked to sing a solo in the choir, how he'd been awarded a form prize for the string of 'A pluses' in English compositions, she'd clapped her hands. He'd sung one of the songs he'd composed for her, the one about the horse with no tail. Her giggles had made his heart sing. They'd stayed on the street for a while, stopping by an electronics shop to listen to Chopsticks' latest song and the soundtrack of *The Private Eyes*, Sam Hui's new movie.

But at the front door, Mama had told him that Stepfather wasn't feeling well, she was nursing him, so they wouldn't be able to see the elephant at Lai Chi Kok Amusement Park the next day after all. Inside, the chandeliers were still filthy dirty and the green patch on the wall had spread to the ceiling. The whole flat smelt mouldier than ever. Overhearing Joey's 'Pooh',

Stepfather had yelled that Joey should feel lucky that there wasn't a water shortage, like in 1967.

At least Mama had bought a videocassette of Roman's latest album for him. It was the first time Joey had seen one. It looked like a black book. It was on the shelf above a new machine called a cassette player, which was linked to the TV. Joey studied the cassette cover. Roman Tam's happy face smiled back at him. Joey pushed it into the player and placed the headphones over his ears.

The machine whirred and Joey entered a vast black cave on the screen where Roman's fans, hundreds of them, were filling every nook and cranny. Electric guitars were tuning, keyboards playing, drums rolling. As the countdown commenced, there was the thrill of flashing lights and the fog of dry ice. And there was Roman standing in the wings, waiting for his cue. As he walked on stage, the fans roared and waved their banners. Roman's costume gleamed and dazzled in the spotlight and the music flowed like water.

'Roman! Roman!' chanted the pretty ladies, and a sexy saxophonist blew clouds of melody across the stage. The violins were singing too, soaring above Roman's melody. A drummer with long hair and earrings beat rhythmic heartbeats.

'Roman! Roman!' Roman's adorers chanted. And Roman snapped his fingers and stamped his feet and a flurry of dancers strutted toward him from the four corners of the stage. Flapping their feathery wings, they stomped their feet to the boom of the beat.

Roman was singing of joy and the sunny south. He sang of machines and dreams and mechanical things. He sang of bitterness and a north wind blowing. He sang of love, sending it deep. And, all the while, voices chanted, 'Roman! Roman!'

A shoulder shake. It was Mama. She sat beside Joey on the sofa, her fragrant long hair still wet from showering. 'He's lovely, isn't he?' she said.

Joey disconnected the headphones. 'Can we listen together?'

Mama sighed. 'Better not.'

Joey humphed, crossing his arms against his chest. 'I *am* going to be a singer,' he said firmly. 'Whatever Stepfather thinks.'

Mama reached for a hand. It was silky smooth. Joey squeezed her forearm. 'Ow!' she said, moving away.

Alarm bells rang in Joey's head. 'He hasn't . . . hit you, has he?'

Mama lowered her head.

'Does he?' said Joey angrily.

'No, no, it isn't like that,' she said. She took Joey's other hand. Her palm felt sweaty. 'Just promise me you won't play the stock market,' she murmured.

Joey rested his head on her shoulder. If Stepfather was hitting Mama, Joey would have to start saving money, fast. Had Stepfather lost his fortune? That would explain why the flat hadn't been decorated. Stepfather had stopped boasting that he'd bought it outright, as well as his Morris Minor. Is that why didn't he drive it to work anymore? Apparently, he didn't even catch the cross-harbour bus. Was that because it was one dollar, rather than ten cents from Blake Pier? Was he still the chairman of the Industrial Plastics Association? An honoured guest of Tiger Association dinners? Questions flooded Joey's head.

'I don't want to talk about him,' Mama said, as if reading his thoughts.

Joey's brain was seething. 'Okay, what about Papa then?' he said. Last night he'd discovered that the cupboard had been completely cleared. He took a sharp intake of breath. 'Where are his Chinese opera costumes?'

'Gone,' said his mama, simply.

Joey blinked in disbelief. 'What? Everything?'

Joey felt her stiffen. Then she leaned forward and whispered in his ear. 'I stored a couple of his best under your mattress.'

But Joey still felt mad. He glanced at the TV to try to calm down. Arms akimbo, Roman was gesturing towards heaven, his round face fixed in a beatific pose.

'That's enough now, Joey. Don't make me feel sad,' said Mama. Her hand had gone limp. She stood up. 'Just remember that your papa was the kindest, most generous papa a woman could ever hope for.'

'What's wrong, Mama? Why are you crying, Mama?'

Mama reached for a tissue. 'Because he died, that's all.'

CHAPTER 18

Bad Dreams

THAT NIGHT JOEY had a nightmare. He was on stage, about to sing. In front of him, a sea of heads bobbed to the music like buoys on an ocean. Behind, the orchestral players rolled in harmony with the waves of Maestro's baton. Joey felt a breeze blowing on his face, his neck, his stomach. He looked down and discovered he was naked. A spotlight above flickered and faded. Then the air felt thin, too thin and Joey was rising, higher, higher, and the music was becoming softer, too soft. Joey strained to hear the oboe melody that was his cue to start singing but its shape was unfamiliar now and in the wrong key. In desperation, he tried to envisage the notes patterning the page but they blurred into sheaves of empty staves. He turned to see Maestro shaking his fist. Ray giggled. A So guffawed. Then there was row upon row of rippling, laughing heads mocking him. The laughter spread and swirled like seaweed, clogging his throat. Joey couldn't sing, even if he wanted to.

Joey awoke to the clanking of a passing street tram, and a beating heart. He searched for what he'd been dreaming about, remembered, cleared his throat and took a deep breath. *La-la-la la.* His voice echoed around the bare walls of his bedroom. Silly him, of course he could still sing. The dream was nonsense. But it scared him all the same. Because if he couldn't sing, he'd be expelled from the school and would have to stay here.

Joey dispelled the nightmare from his mind by making a plan for the day. Mama had persuaded Stepfather to let him visit Aunty Tam without her. Joey would make some pocket money by selling a few oranges there. He stepped out of bed to open the blinds.

Mist hung low as it often does in April. It would clear into a bright blue day of heat and humidity. The Asian cuckoo was calling. Mr Waters had taught Joey to recognise a few birds. He'd told him that one of his favourite places in Hong Kong was the Bird Market, that during the holiday he'd visit Mai Po marshes where you could sit in the middle of mudflats and spot numerous species through binoculars. Joey would ask his mama to buy him a pair.

'I'm cooking your breakfast,' Mama called from the kitchen. She would be frying an egg to top his breakfast noodles.

'Where's mine?' called Stepfather, hawking and spitting. Joey couldn't wait to get away from that sound.

Sparky's head was poking out from under the orange stall and he wagged his little tail like a feather duster upon seeing Joey. Joey knelt down to stroke him and received a faceful of licks.

'He and Mei Mei are firm friends now!' said Aunty Tam.

Joey reached for the hosepipe. While washing off the doggy drool, he heard how Uncle Bo now liked Sparky too, because all their food scraps were eaten. The three of them slept together in a double bed.

'Good boy,' said Joey, stroking Sparky's velvety ears.

Uncle Bo appeared with a basket of laundry and started hanging the T-shirts on coat hangers. 'Have you eaten?' he asked.

Aunty Tam reached for a watermelon, knocking the rind with her knuckles to check the fruit's ripeness. With a cleaver, she halved it and sliced through its juicy flesh. Then the three of them sat on stools behind the counter to eat it, spitting out the seeds in the gutter. It wasn't the sweetest Joey had eaten, by far. And somehow they got talking about Stepfather.

'He wants you to get rich so you can pay him some pocket money, put food on the table, marry a pretty lady,' explained Aunty Tam.

Uncle Bo nodded.

'I *will* earn a lot of money,' said Joey hotly, 'when I become famous.'

Aunty Tam tightened her lips to mimic him. ' "When I become famous." '

Uncle Bo spat out a mouthful of seeds. 'Easier said than done!'

Joey felt blood rushing to his cheeks. 'Until then, I'm going to save money by selling oranges with you,' he said.

Aunty Tam wiped her mouth with the back of her hand, her watermelon rind mimicking her mocking smile. Joey threw his into the bin. 'In the holidays. . . .' he continued, his voice tailing off. He looked up at his aunty, her flat nose, her piggy eyes, the pores dimpling her chubby cheeks. Did she really think the only way to be successful was to set up a business, be a doctor or a lawyer or an accountant? He could barely hide his disappointment.

Aunty Tam's face softened. 'Joey, of course we don't mind you working here,' she said. 'You bring us luck, Orange King, never doubt that. But I'm afraid you'll never earn very much.'

Uncle Bo grimaced.

'Why should I listen to Stepfather anyway?' Joey retorted. 'He's not my real—'

'But you *must* listen to him,' said Uncle Bo sternly.

Anger prickled Joey's skin. He blamed Confucius rather than his uncle, whom he liked. Because Confucian values demanded children obey their parents. But did that mean Joey should obey his stepfather? Even though the monster beat him and didn't consider him his son, they weren't related by blood. Come to think of it, Uncle Bo wasn't a blood relative either, so why should he listen to him? Besides, Stepfather didn't respect Uncle Bo and Uncle Bo didn't give face to Stepfather. Last Chinese New Year for example, he hadn't even come to Happy Valley with Aunty Tam to *'baai nin'*.

Joey had only one line of defence left. He filled his lungs and raised his head, saying, 'I sing because my mother likes it.'

'In the shower,' said Aunty Tam.

Joey felt his body tensing again. Surely it was clear that his mama encouraged him, wasn't it? Whenever he sang to her, she closed her eyes and smiled. Was she pretending? 'Canto stars make a lot of money,' he mumbled.

Aunty Tam sighed. 'Next time I visit Man Mo temple, I'll light some incense for you.'

Uncle Bo stood up and directed his gaze towards Aunty Tam. 'You think burning a few joss sticks would help?'

Joey wondered too. He hunched his shoulders and stroked Sparky. Why shouldn't Joey sing to make people happy rather than work in a factory or office? Wasn't spreading happiness more precious than making money? What was the point of having factories and cars and expensive flats if you were miserable? Uncle Bo and Aunty Tam were poor but they really cared for each other. Joey had even seen them kiss, once.

Aunty Tam reached for Joey's hand. 'Let's wait and see what the gods have in store for you,' she said. 'Let's wait and see.'

CHAPTER 19

Thieving Again

IT WAS EASTER MONDAY, the day Jesus rose to heaven. Joey knew the story because he'd sung hymns in the school choir about the resurrection. Three days before, Good Friday, King Herod had tricked people into believing that Jesus was a bad guy and Jesus had been nailed to a cross in his underpants. Just before he died, he'd called, 'Eli, Eli, why hast thou forsaken me?' Who Eli was, Joey wasn't quite sure, but he guessed it was God. But God hadn't forsaken his son because three days later, like magic, Jesus disappeared from his coffin and flew up to heaven. Joey looked up to the sky imagining Jesus flying by. Washing flapped in the wind on poles poking from the balcony above. In naughtier days with Todd and Shrimp, they'd used long poles to hook lady's bras.

Yuck! Stepfather was smoking in the living room. Joey pretended to collapse in a heap of coughing.

Mama lifted her head from the movie magazine she was leafing through. 'Behave yourself,' she said.

'Have you ever been to church, Mama?' Joey asked.

Mama cupped her chin with her hand and thought for a while. 'I suppose so,' she said. 'Because I've been to temples.'

Temples were Chinese churches. Smoky little places where people burned joss sticks and beseeched gods for good fortune. Were those gods Jesus in another form? If they weren't, who were they? And why did people believe in them? Joey hadn't thought about this before. 'Who do you worship?' he asked.

Mama glanced at an image of Nancy Sit before answering. 'Guan Yin, the goddess of mercy,' she said. 'But there are many others.'

Stepfather stubbed out his cigarette. 'That's enough questions, Joey. I've booked at the Luk Yu.'

Wah! Going to a *dim sum* restaurant was a treat. Stepfather must have won at the races. Joey felt hungry as a horse: dumplings in soup, barbecued pork, chicken wings, beef brisket noodles. He crammed his stomach but there was still room for more. On the way back, Stepfather grudgingly gave him a few coins to buy a bun at the bakery.

On the street, Joey bumped into Todd. He'd grown so tall that Joey didn't recognise him at first. He was wearing a vest and sunglasses and his voice had broken. 'Shrimp's buying drinks,' he said, jutting his chin towards a stall opposite.

Joey hadn't told either of them that he was back home. He'd secretly hoped their paths wouldn't cross but Happy Valley Racecourse was a few minutes' walk away. Shrimp waved. They walked to a nearby playground and drank Coca-Cola on a wall.

'Wanna see my dragon tattoo?' Todd yanked the back of his T-shirt. *Wow!* A green dragon snaked across his shoulder blades.

'It didn't hurt too much, right?' said Joey.

Todd gave him a withering look.

Shrimp had one too, on his forearm. Sam's was on his calf. A sixth-former had taken them to a parlour in Mong Kok.

'Who's Sam?' Joey felt a tinge of jealousy. Ah yes, a boy from the Mainland who'd joined his old school shortly before he left.

Todd adjusted his sunglasses. 'Hey Joey,' he said. 'I've found a great place to lift some stuff.'

'Oh yeah?'

Todd's teeth gleamed. 'The temple at the end of this street,' he said. 'There's an old caretaker who minds it but he often slips out for a pee, or tea.'

Todd's plan was to raid the temple when the old man wasn't there. One of them would stand guard and warn the other two if anyone entered. Fat green grass snakes squirmed around Joey's stomach. Weren't temples, even Chinese ones, holy places?

Shrimp didn't look too happy about the idea either. 'That's where my grandma prays I'll pass my exams,' he said.

'No chance of that,' scoffed Todd.

Shrimp's voice dropped to an undertone. 'She says she sometimes sees my great-grandfather's ghost there.'

Joey shuddered. 'How come?'

'He was brought back there,' replied Shrimp. 'From America. Because it's bad luck to be buried so far away.'

'Ugh,' said Joey, imagining a hull stuffed full of dead railroad workers crossing the Pacific.

'I helped Grandma and my aunties fold hell money there for his funeral,' Shrimp continued.

Todd flinched.

Joey had attended a Taoist ceremony for Stepfather's brother there too. After priests had lit joss sticks and chanted prayers, the paper-mache washing machines, fridges and radios, even a car, had been burnt to ashes. Joey slurped the last mouthful of Coca-Cola. 'It's not right,' he said. 'I won't join.'

'Scaredy-cat,' said Todd, *kung fu* kicking the wall.

'I'm not scared,' said Joey hotly, 'It's just. . . .'

They were all standing now.

'Maybe we just go for the alms box?' said Shrimp.

'Or the money in the drawer,' said Todd, 'if we can force the lock.'

Joey scrunched his Coca-Cola tin. 'Can't be that easy,' he said.

Todd scowled. 'Meet you at 10 am sharp tomorrow.'

'I'll only come if I can be the lookout,' Joey said, as firmly as he dared.

'See you tomorrow,' snapped Todd.

Joey regretted going to the temple the moment he arrived. He stood underneath a frieze of dragons on the green-tiled roof and peered into the gloomy interior. Two door gods glared back at him. They were clutching spears and showing the whites of their eyes. Red-eyed dragons were stuccoed on columns, like giant geckos.

He couldn't see much of the interior because smoke from coils of sandalwood clouded the air but a beam of light brightened one of the shrines. The god was sitting cross-legged behind glass. In front of him there were fruit and flower offerings, and spiky crew-cuts of burning joss sticks littering a tray. Many people must have visited earlier to pay their respects.

To Joey's left, an ancient man with a long white beard and thick-rimmed glasses. The caretaker was studying a thick tome while twirling lucky sticks in a tin box. On the table there was incense, candles and talismans, as well as the alms box. Joey guessed the drawer Todd had mentioned was underneath.

Joey took a deep breath and stepped over the door stone. He'd walk around once then wait on the street. Instinctively, he looked for places he could hide if a visitor entered while they were thieving. There was a huge bronze bell but the taut hide skins of both ends were firmly nailed down. Opposite, a scarlet and yellow drum was chained to a display cabinet. He could hide behind it, or climb up to lie across the rafter which bisected the temple. He leaned on the drum to see if it would bear his weight. It could! On a central table were a collection of porcelain plates. They must be worth something. Some small statues too. But what was worth stealing? Just the money, Joey supposed. How much could there be inside an alms box? Wasn't it for the Chinese gods? What if the gods were real and came down from heaven to curse them?

Todd and Shrimp appeared at the entrance, casting sinister shadows. Joey sauntered towards them pretending to be okay but inside he felt a gnawing sense of something nasty consuming him. He followed Todd to a rubbish bin and they sat on a wall behind it to wait.

Eleven o'clock and the caretaker still hadn't emerged from the temple. Only a few old ladies had entered. A fly buzzed by. 'Do you think the geezer's ever leaving?' Joey said.

Todd cursed.

Shrimp checked his watch.

No, this is just wrong, thought Joey. He stood up. 'I need to go home,' he said.

'I'm staying here,' Todd growled, looking over to Shrimp.

Shrimp cleared his throat and gave Joey a guilty look. 'Me too.'

'See you then,' said Joey, as flippantly as he could.

'Loser,' growled Todd.

Joey set off down the street, swaggering from side to side, trying to look as if he didn't care. But the more he thought about it, the more he really didn't want to start thieving with Todd and Shrimp again. He'd much rather be at home watching that Roman Tam videocassette, or listening to some Cantopop. Stealing from a sacred space made him feel like a green-blooded vampire, and the risk of getting caught was real.

Most of all, he longed to get back to school.

CHAPTER 20

Horrible Histories

IT WAS GREAT TO BE BACK, although Joey's school shoes pinched his toes. Mama had bought them only a few months ago but there was no way he would ask her to beg Stepfather for money to buy a bigger pair. Joey sellotaped a poster about Roman Tam's forthcoming concert series on the wall above his bed and went downstairs.

The common room was abuzz with talk about a big fund-raising concert at the end of the term. There would be auditions for solo parts in a couple of weeks.

Joey's new bed in the dormitory had a direct view of Lion Rock. After lights out, Joey sneaked into the bathroom with Ray, Ben and A So for a game of stone, paper, scissors. Ben showed off his collection of car cards, A So shared some barbecued cuttlefish his grandma had smoked. Ray traded pictures of Chinese instruments for their Chinese History project. They played a guessing game in which Joey imitated musical instruments with his voice. His *suona* sound had his friends sniggering with laughter. Then they joked about Miss Wu, wishing she'd wear more revealing Western clothes rather than just *cheong sams*.

The next morning Joey awoke at daybreak and opened the window. The birds were chirping and a fresh breeze flowed down from Lion Rock mountain, its strength and majesty filling him with pride at being a pupil at such a fantastic school. Now what did he need to pack in his satchel? He'd memorised last term's timetable but this term's had changed. There were more choir practices and singing lessons. And, first lesson on that day, a brand new subject: Music History. Joey smelt his fresh new notebook, placed it inside his satchel and skipped down to breakfast.

Room 702. As usual, Joey sat between Ray and A So. After everyone had quietened down, the teacher, Mr Clarke, introduced himself. The old white man looked as ancient as Confucius, with curly grey hair and matching moustache. Already perspiring, he took off his tweed jacket and turned up the fan. A So tittered, pointing to the jacket. It was sprinkled with dandruff.

Mr Clarke turned and smiled. 'I look forward to getting to know you all individually,' he said, 'but for now I want you to close your eyes. Because we're going to go back in time.'

Joey glanced around his classmates to see whether they were following the instruction.

'I mean it,' said Mr Clarke firmly.

The whirr of the fan, the buzz of a bee. The silence was tangible.

'I want you to imagine the beginning of time,' said Mr Clarke, 'a time when there wasn't music as we know it, only the hum of the universe.'

Ben chuckled quietly. Joey peeped to observe Mr Clarke's reaction but he'd chosen to ignore it. Or he was deaf. The teacher's eyes were closed, his arms extended as if invoking the gargantuan hum.

'Imagine,' continued Mr Clarke, 'you're among the first on Earth, the first *Homo sapiens,* living in a land where the only sounds were the whispering of the wind, the growl of thunder, a pig roasting on a spit. Melody was birdsong, a rooster's cry. Rhythm was the *whoosh* of the waves, the *tap tap* of your father's stone tool, the pitter-patter of rain. Harmony was sitting round an open fire, picking nits from your sister's hair, sharing stories, studying the stars.'

Joey took a deep breath, He enjoyed creating pictures in his head.

'Over the millennia, societies grew,' said Mr Clarke. 'From clans and hamlets to villages and towns. People became civilised by having to learn to live together in cities. Organised religion began, and an appreciation of art. Western music was born in monasteries and churches. It was sung by priests. At first, a single priest, who, obviously, sang a single melody, a melody we now call plainsong. Then two priests, singing two melodies at the same time, which we now call counterpoint.'

So many new words. Joey vowed to study them.

The priests' prayers became longer and more complicated and it became necessary for music to be written down. That's when composers came into existence. They devised a new language, a new method of notating sounds. Then new instruments were invented – violins, fortepianos, clarinets, horns – inspiring composers to write music for them. Until the sixteenth century, the main purpose of making music was to praise the Lord.'

'Amen,' whispered A So.

'*Sshhh,*' hissed Joey, irritated. He believed that composers had magical powers and he wanted to learn more about them.

Silence, except for the very distant hum of an approaching aeroplane. Joey peeked and Mr Clarke gave him a fleeting smile. 'You can open your eyes now,' he said. Joey blinked.

Mr Clarke was rolling a piece of chalk in his hand and looking pensive. He motioned towards the blackboard and wrote new vocabulary on it. Joey squeezed the cartridge of his pen. In his best handwriting, he carefully copied the words, writing them three times to aid his memory.

Before bedtime, Joey usually sat in the common room to chat with friends. Television was only allowed on special occasions and radio was only for the evening news, which Joey rarely listened to because it was boring. While making cocoa, his ears pricked up at a conversation Maestro and Miss Asimov were having. They were telling some older pupils about their life in Shanghai during the darkest days of the Cultural Revolution. They used to listen to recordings of music under beds after the curtains were shut and the lights were out because Western music was banned. Red Guards would break into homes at any time to smash the records and hurt people. It was the first time Joey had heard about foreigners needing to escape from China.

'Why did you come, to Hong Kong, I mean?' he asked.

Maestro took a sip from his cup of tea. 'Because we could. Britain and Russia were allies in World War II. Miss Asimov and I had already left Russia and we weren't welcome back. Later, we weren't welcome in Shanghai either.' Maestro paused for breath. 'Were you born here, Joey?'

Joey shook his head. He told them all he knew about his mama carrying him across the border after his papa died.

'What year was that?' Maestro asked.

Joey blushed. He wasn't sure. 'My stepfather came in 1962,' he said.

Maestro looked solemn. 'That was the year the border between Hong Kong and Guangdong was opened for three days,' he said.

'We were all so poor in those days, weren't we, Father?' said Miss Asimov.

The clink of Maestro's china cup added weight to his answer. 'We were indeed,' he said, 'but we kept our dignity, didn't we. Looking back, that was because the Jewish community here were so kind to us. They helped us find work.'

What was 'dignity'? Joey would look up that word in the dictionary.

'How long did we live in that leaky shack on Lion Rock?' said Miss Asimov. 'The place where we'd be woken by the early morning squeals of pigs being slaughtered?'

Joey squirmed at the memory of that sound.

Maestro couldn't remember exactly. 'What I do recall is how hard Chinese immigrants worked. Many would have two of three jobs at the same time. But everyone was so hopeful, so helpful, in such good spirits.' He turned to Joey and looked him straight in the eye. 'We have a lot to learn from you Cantonese people,' he said. 'Tenacity is in your blood.'

Joey would have to look up that word too, the one that began with T.

Miss Asimov finished her cream bun. She had a dreamy look in her eyes as she reminisced about their home in Shanghai, 'I remember the bubbling samovar in our kitchen and the smell of Mamushka's freshly made bread and jam,' she said.

'What about the lashings of thick double cream I'd pile on when you performed well?' asked Maestro, grinning broadly. One of his front teeth had fallen out. Apparently, he had taught her the piano and although she practised really hard, she didn't have the talent to be a soloist. But she'd always earned money from teaching the instrument. I'll be a teacher too, thought Joey, if I can't make it as a performer. Miss Asimov could teach piano in Hong Kong because she could speak the Cantonese she'd learned at school. He could teach in many countries around the world if his English was good.

Before bed, Joey queued for the phone to call his mama. He had a few questions to ask her about his early life. He told her he had no interest in visiting the Mainland, apart from paying respects at his papa's grave in Jiangsu one day. And he mentioned how the Asimovs had been penniless, like them, when they had crossed the border from the Mainland.

'Are you sure?' Mama said.

CHAPTER 21

Joey the Honest Boy

L IKE MOST DAYS, Joey had borrowed an English book from the library overnight. He'd chosen from a special display entitled Famous Asians. Yo-Yo Ma was a Chinese musician who lived in America. Joey had read the biography from cover to cover three times before sleep. Yo-Yo was one of the finest cello players in the world. He'd asked for an instrument on his third birthday. By seven, he'd already played to the American president's wife and five thousand people. In every illustration, Yo-Yo was smiling. 'No pain, no gain,' was his motto. He wasn't sad that

his father made him practise for hours each day. Because he
was doing something he loved, Joey thought.

Joey's first class was double English. He arrived early at the
library, the janitor unlocked the door for him. First, he checked
the meanings of the words 'dignity' and 'tenacity'. He liked
using dictionaries to learn new vocabulary and, for some
reason, could memorise a new English word immediately.
Then he took the borrowed book out of his satchel. *Eiya!* It
was wet. Its inside pages were sopping. His water bottle had
tipped upside down.

Joey tried to separate the pages that had stuck together. He
had to return it that morning. *Driiiing!* The bell made him
jump. How could he put the book back on the shelf in this
state? He heard approaching footsteps, the sound of voices. His
classmates were coming. Joey quickly spread the book out like
a fan, balancing it upright in its allotted space. A sixth-former
entered with an armful of books. Joey pretended to peruse the
History section until Maisy came.

'Good morning, class,' said Miss Williams when everyone
was assembled around the circular table. The thick lens of her
glasses made her eyes the size of peanuts. She was the oldest
teacher in the school. Ben joked she was half man because of
the dark hairs above her upper lip.

'Joey did the best,' she said, handing back his spelling book.
She'd warned them the test was difficult. Nine out of ten. That
cheered him up. But what a silly language English was! There
were different words for every object and you had to memorise
each one. The Chinese language was so much easier. Characters
gave large hints about their derivation, meaning and pronun-
ciation. 'So, for example,' Joey had once explained to Vasavi,
the only Indian in his class, 'a character which starts with the

branches of a tree has something to do with wood. And a character which has three splashy dashes is linked to water.' Vasavi hadn't looked convinced.

English spelling wasn't consistent either. How could you guess that there isn't an *h* after the *w* in 'withstand'? As for those sneaky consonants that crimped like cocoons – *sp* and *spr*, *ch* and *chr* – Joey still found it a challenge to single them out.

'S . . . sp . . . spr. . . .' he said.

'*Sssh,*' said Maisy, elbowing him.

Miss Williams had marked Creative Writing too. She handed Joey his composition book. A+. But ugh, three red circles! That pesky word 'the' had let him down again. He often dithered about when to use it and it was impossible to pronounce anyway.

'Say it,' said Miss Williams, pointing.

Joey pushed his tongue forward and held it between his teeth. That's where it needed to be to start enunciating the sound correctly. He took a deep breath, exhaled, his lips tickling, spittle bubbling, until – now! He flicked his tongue back like a dragon retreating into a cave.

His Chinese classmates were trying to pronounce *th* too, their dragons spitting with various degrees of venom.

'You're getting there!' said Miss Williams. She voiced the difference between *th* and *f* and *fr* and asked them all to practise for homework.

Driiing! Next period was reading English books, with Miss Vonn. Joey loved her classes. She was young and friendly and wore her hair in fashionable braids. And English books were so much more interesting that Chinese ones.

'Let's read some non-fiction today,' she said, walking over to the special display of books. 'Who's noticed this collection?'

Joey's heart skipped a beat as his classmates buzzed with anticipation about which one Miss Vonn would choose. She picked one on the left-hand side. 'How about Mahatma Gandhi? The famous man who drove the British out of India by peaceful means.'

'Yes, Miss, let's have Gandhi,' said Vasavi.

'You read us Gandhi last time,' said Ben.

'Oh,' said Miss Vonn, covering her mouth with a hand.

'How about someone Chinese?' said Kwok.

'Bruce Lee, Bruce Lee,' chanted A So, flexing his fists in a *kung fu* pose.

'Good idea,' piped Joey.

'*Oooh* yes,' said the Chinese girls.

'*Mmm,*' said Miss Vonn, examining the book covers.

Joey shrank back into his seat. Was she focusing on the photograph of Yo-Yo and his cello? Surely not.

She was. Reaching forward, she selected it. 'This is beautifully told,' she said.

How could it be possible? There were so many others she could have chosen from!

'Yo-Yo Ma, the Cello Star,' she said, reading from the jacket flap. She sat back at the table, swivelling her teacher's chair so everyone could see the cover. Flipping to the back, she started reading from it: 'Yo-Yo Ma, the amazing Asian American. . . .'

Miss Vonn turned to the first page. Joey's heart knocked at his ribs. She hadn't seemed to notice that some pages were stuck together. Maybe they'd already dried. 'Yo-Yo was born in Paris,' she began. 'His parents were very poor. His elder sister played the violin.'

Joey listened in horrified fascination. Every time Miss Vonn turned a page, her hand reached for the top right-hand corner.

'But Yo-Yo wanted to play the—'

She stopped reading. Joey stopped breathing.

'Eeooww,' she said. 'This book is wet.'

She tried to divide the next soggy pages.

Ben tittered.

'Who would be so careless?' Miss Vonn said, looking around the table. His classmates were shrugging their shoulders, shaking their heads, eyeing each other nervously. Blood rushed to Joey's cheeks.

'Was this one of you? This *is* the first class of the day,' said Miss Vonn.

Silence. Sweat beaded on Joey's forehead.

'If no one admits it, I'll have to check the sign-outs,' warned Miss Vonn.

Joey still didn't say a word. Confessing was on the tip of his tongue but something prevented him.

Miss Vonn walked over to the counter. How Joey wished he could crawl under the table and disappear.

The register was open on the latest page. Miss Vonn's index finger slowly moved down the list. 'Goodness me,' she said and a surge of adrenaline raced up Joey's spine. But she walked back, picked the book up and carried it to the window ledge. 'I think Yo-Yo would benefit from some sunshine,' she said.

Bruce Lee. The fighting spirit. Miss Vonn was ignoring Joey. She didn't even glance at him once. She must have read his name in the register. Would she report him to Miss Asimov? He wanted to run. As Bruce Lee's story dragged on, he tried to reassure himself that Miss Asimov would be kind. After all, he hadn't stolen the book, or dumped it, had he? And after it dried out, it would be fine, right? Just a bit crinkly in places.

Driing. Break time. Joey looked up and caught Miss Vonn's eyes. She knew. She asked him to stay behind. Joey's classmates quietly filed out. Maisy glanced back.

'It was me,' Joey blurted, as soon as he was out of earshot.

'Too late,' said Miss Vonn sternly. 'Follow me.'

In the office, Miss Asimov listened carefully. 'Accidents do happen,' she said, 'but that's not the point. Why didn't you own up to Miss Vonn?'

'I did.'

Miss Vonn frowned. 'Not immediately. Do you think that's good enough?'

'And before class, no librarian there tell.' In panic, Joey's brain reverted to pidgin English.

'There wasn't a librarian there to tell,' said Miss Asimov, correcting him. 'And don't be insolent.'

Joey lowered his head. Now he felt even worse. He hated the feeling of being disliked.

Miss Asimov raised her chin. 'I suppose you will learn something from this,' she said.

Joey swallowed but the lump of sadness wouldn't go away.

Miss Vonn said she had to go to prepare for her next lesson and Miss Asimov closed the door behind her. 'Joey?' she said. 'Joey, look at me.'

Their eyes met. Hers were big and blue and shining. Did his reveal all the lies he'd told? All the stories? Could she see all the money he'd pickpocketed? Were they written on his face? My stepfather beats me, he'd say, in defence. But maybe that was still good reason for her to dislike him forever.

'Joey you are an honest boy, aren't you?' she said.

Words gushed up Joey's throat but shame suppressed them. He felt unclean, like a dirty rag, compared to such a pure lady.

Miss Asimov rested a hand on his shoulder. 'I have a suggestion,' she said calmly. 'Whenever you find yourself being untruthful, stop, find a peaceful place, and give yourself some time to reflect.'

'Okay, Miss, I will,' said Joey, unsure of what she meant precisely but responding to her care. 'And starting from tomorrow,' he said, 'I will pinch myself every time . . . I could be more honest,' he said.

Miss Asimov looked as if she'd just seen a bowl of cherries and fresh Russian cream. 'That sounds a splendid idea,' she said. 'Good for you!'

It was the first time Joey had heard about reflection. That night, after looking it up in the dictionary, he thought he saw the merit of it.

CHAPTER 22

Dissonance

'N EXT,' DRAWLED MR LEACH'S VOICE from practice room 34.
A black pianist, the best in school, came out, and Joey
entered.

'For the solo singing spot, right?' asked the Head of Music.

'Yes, sir,' replied Joey, expanding his chest.

Auditions for two soloists for the fundraising concert. Both
would be accompanied by full orchestra. Joey had put his name
down as soon as he'd heard.

'Music, please,' said Mr Leach, gesturing from the piano.

'It's in here,' Joey replied, tapping his head.

Mr Leach scratched the bald patch on his head. 'You didn't think I would need a copy?'

'I can write it out for you. Tonight.'

Mr Leach blinked. 'What's it called?'

'*Under Lion Rock.* It's a Chinese song.'

'Well, I'll be damned,' replied Mr Leach, and Joey flinched. Wasn't damned a swear word?

'I climbed Lion Rock only last weekend,' the teacher continued. 'Terrific view from the lion's head. Okay, young man, sing me the melody.'

At the end, Mr Leach clapped. Then Joey told him that Mr Downs had offered to help him transcribe the accompaniment for Western orchestra.

'Transcriptions of Chinese instruments? Awesome, as Mr Downs would say,' said Mr Leach.

'I will do it, sir,' said Joey.

As Joey passed the school office on the way to his next class, he heard the raised voices of Maestro and Miss Asimov. Arguments always unsettled him. He leant towards the door and cocked his ear.

'Calm down, dear,' Maestro was saying.

'I can't believe that such a rich man didn't have the intelligence to make a will,' wailed Miss Asimov.

'But Chinese people believe it tempts fate to write one.'

'Surely he intended to keep supporting our pupils after he died.'

Joey leant closer. What was that about supporting the school? He supposed the money had to come from somewhere. And, come to think of it, what about his scholarship?

'How ludicrous to believe in bad luck in this modern age,' continued Miss Asimov. 'It's not as if Mr Lee didn't know we're already finding it hard to pay our bills.'

Who was Mr Lee? A major donor, Joey supposed.

Maestro's voice lowered to a normal volume. 'How about you try to respect their culture more?' he said. 'The Chinese people have offered us a home here.'

Miss Asimov's was still shrill. 'Talk me through what you think will happen,' she said. 'Should we carry on as normal, until the landlord is knocking and the teachers are suing for wages?'

Joey stepped backwards to digest what he'd heard.

'How about you call the landlord?' Maestro said. 'Arrange a meeting. Use your charm to try to persuade him to lower the rent. Oh please don't cry, my dear.'

Silence, apart from Miss Asimov blowing her nose. 'If we close at the end of next month, we'll have to cancel our fundraising concert,' she said miserably.

The fundraising concert? Oh no!

'Tell you what,' continued Maestro. 'Robin's grandson has his bar mitzvah this Saturday. Let me ask him whether he knows of anyone who would be willing to donate more.'

'Joey?' Mr Clarke was calling him. 'We're waiting for you.'

Joey found a desk near the back and slumped over it. Music History, hot and sticky. An aeroplane grumbled overhead, rattling the pots of Miss Asimov's potted palms. Joey's mind was reeling. The school might have to close? How terrible was that? He'd have to go back to Saint Thomas's. Live in Happy Valley again.

Maisy poked him with a ruler from behind. 'Are you okay?' she mouthed.

Joey smiled sadly.

Mr Clarke was taking his tweed jacket off. There were dark patches on his shirt under his armpits. 'Today's new musical word is dissonance. D-I-S-S-O-N-A-N-C-E,' he said.

The evil hisses of the double S made Joey shiver.

'A lot of contemporary music is dissonant,' said Mr Clarke, 'portraying the conflict, disharmony and disorder of our times.'

Joey blocked his ears when Mr Clarke played a sequence of ugly chords on the piano. Dissonance, to him, was Miss Asimov's crying. Maestro's pleading. What could he do to help?

Slap! Joey jumped with the smack of Mr Clarke's ruler on his desk. 'You can unblock your ears now,' said his teacher.

'Sorry, sir,' said Joey. His classmates were laughing but he didn't care.

Mr Clarke was playing some harmonious chords on the piano. C major, G major, Bb minor: Joey recognised them instantly. Having perfect pitch made it easy.

'The opposite of dissonance is consonance,' continued Mr Clarke, 'but within a work of art, for consonance to sound its most beautiful, we've had to have experienced dissonance also. Think of any piece of Mozart's.'

Sharp as a knife, Mr Clarke sliced the note D into a quiet C major chord. The D longed to be resolved, like Joey's thoughts. His heart quivered on the blade until that D sunk into a C. And the moment it did, an idea floated into his head. He didn't have Sparky to help him but there were other ways – in a novel he'd read about some boys in America. He would re-read that chapter, stop, and sleep on it.

Five minutes before lights out Ray was hiding under his duvet snivelling. The scroll of his violin peeped out like a man with a long face, a curled wig and pegs that tuned his brain.

'*Sssh,* I can't get to sleep,' said A So.

'Me too,' said Ben.

Miss Asimov did her final round and then the four of them crept to the bathroom. They sat in the dark listening to the sound of a dripping tap and the *ZZZZZ* of a mosquito. Joey switched the fluorescent light on to find it, and squashed it on the mirror. 'No need to jiggle legs anymore,' he said, switching off the light.

Moonlight cast shadows over the space revealing that Ray was still upset about something. 'I'm not pretending,' he said. 'I must have practised that passage a thousand times, but I still fluff it up.'

Joey knew the passage only too well. Everyone knew it. Whether Ray was in a practice room or the dormitory, in an orchestral rehearsal or playing chamber music, he'd be practising one particular phrase where his hand had to jump up the fingerboard as quick as a karate kick. Whenever Ray couldn't practise, like in class, he'd be twisting his left forearm and drumming his fingers on an imaginary violin.

Joey told him about a section in a piano piece he'd also not mastered. He could play the right hand perfectly. From beginning to end. No stopping. But when he tried to add the left, his pesky second finger would always forget the B flat.

'But you're lucky to be a singer,' said Ray. 'All you have to do is open your mouth.'

CHAPTER 23

Touching Fame

LION ROCK LOOMED ABOVE the children, its flanks blotched with squatter huts and makeshift shelters. 'Not much further!' called Joey striding up the road with Ray, Ben, Maisy and A So. His arms ached with the weight of the music stands, stools and music on his back but his heart was brimming. He, and Ray, had been selected as the soloists in the fundraising concert! Mr Waters had also told him that Mr Leach had suggested an application for a music scholarship in England.

Also, his friends had agreed to this plan! Ray worried that the sunshine may damage his precious violin but the other three were enthusiastic. Joey turned to give them an encouraging wave. Back of the line was A So swinging the placard Maisy had cobbled together with a broomstick and cardboard earlier that morning. SUPPORT SASSOON SCHOOL OF MUSIC, it read. When curious pedestrians pointed at it, Joey doffed his cap and shouted, 'Follow us!'

Busking. That's what this was called. In New York, musicians played outside subway stations or on busy thoroughfares. Joey had chosen Golden Harvest movie studio on Diamond Hill. There were already many people milling around the area. They were hoping to catch a glimpse of someone famous, Joey supposed. He chose to perform in front of the studio's portico, opening the music stands and clipping the musical arrangements he'd hastily composed. *Whoosh!* The slipstream of a passing car blew Maisy's sheets away while she was rosining her bow.

Those interested were already loitering and circling around them. Joey straightened his tie and placed a plastic bucket on the pavement. *Tink!* A young woman dropped a coin into it. There was a final positioning of stands, a last tuning of instruments and they were ready to go. Joey cleared his throat, ran his tongue along his teeth and licked his lips. He began with a lively number, accompanied by the instrumentalists. *Tink, tink!* A sprinkle of coins hit the bucket. *Tink, tink!* The coins were coming thick and fast. Making money this way was much quicker than selling oranges!

'Bravo!' called a man in a sombrero hat.

A group of Buddhist nuns clapped politely.

Joey chose *Wait For Me*, another song with a jazzy rhythm. He tapped one foot and swung his hips. A man in a T-shirt and shorts dropped a brown banknote towards the bucket. Five dollars! A rush of air danced it along the road. Joey jumped sideways to stamp on it.

'Good job, Joey,' called Maisy.

Joey blushed. It felt good to please her. Ray and Ben were smiling too. A So was eyeing up a girl in a miniskirt and curled hair. There were many pretty girls around and more and more were amassing. Boy, was this fun! Joey raised his head, gestured towards the splendid mountain above and sang the opening strains of *Under Lion Rock*.

> *Life has its joys*
> *But often has sorrows too.*

A well-dressed *tai tai* withdrew an embroidered handkerchief from her handbag and wiped her eyes with it. A couple of passing Filipino musicians carrying guitar cases waved. One day, thought Joey, I'll get a job performing at a tea dance, like them.

> *When we all meet under the Lion's Rock*
> *Our laughter exceeds our sighs.*

'Sure!' called a delivery boy on a bicycle.

Ugh oh. A couple of policemen on the beat were walking towards them. Was busking illegal? Surely not. Joey lowered his voice.

Putting aside our hearts' conflicts
Together we pursue our dreams.

The policeman with a baton stopped, cocked his head and slapped his thigh. 'My favourite,' he said. The other leaned against a nearby wall and lit a cigarette. Joey filled his lungs and sang in their direction.

In the same boat we promise to go together
Without doubt or fear.

The policeman leaning against the wall was singing along too.

Putting aside our hearts' conflicts
Together we pursue our dreams.

'More, more,' people called when Joey finished the song. 'No tips,' warned the policemen with the baton.

Next, Joey sang a happy ditty that lit up his heart like firecrackers. *Tink tink.* Despite the policeman's order, the large crowd was donating generously. And the policemen were still there, jigging their heads to the beat. At the end of the song, they walked over to a group of policemen assembling under a banyan tree on the opposite side of the road.

A large grey cloud doffed the sun's hat and Joey sang a sad song. *Tink, tink!* Something was happening behind Joey's back, diverting the attention of his audience. There was the click of cameras and a babble of excited voices. Joey looked over his shoulder. Stepping down the stone stairs of the production company was a Chinese man in a white double-breasted sailor's suit. Not tall, but very handsome, tailed by two beefy bodyguards.

Joey couldn't believe his eyes. He would recognise him anywhere. It was Roman Tam!

The crowd surged towards the portico. 'Roman! Roman!' they chanted. Then the policemen were cordoning people off, blowing whistles, barking down walkie-talkies.

Roman stopped at a column and tipped his sailor's cap. He was looking over in Joey's direction. As if at him!

'Keep going!' Joey cried to his classmates who were gaping at the commotion instead of playing.

'From letter D,' called Ray, picking up from where they'd left off.

'Repeat!' said Joey, when they reached the end of the song. He kept singing by humming through the introduction, continuously fixing his eyes on Roman, willing him to come over.

And he did! Brushing aside a wiry man, with the bodyguards stumbling to keep up, he strode towards them. Joey smelled a whiff of cologne. 'You've a good voice, little friend,' he said to Joey.

Joey ballooned with pride. At that moment, he wanted to sing forever.

Roman had a gentleness to him. He raised his arm, pointed at A So's placard. 'Sassoon School of Music?'

'It's our school,' replied Joey breathlessly. 'It's run out of money. It might have to close.'

His classmates nodded.

Roman frowned. 'Never heard of it. Where is it?'

Joey swallowed, resisting the instinct to say it was just around the corner. 'Kowloon Tong.'

'Really?' replied Roman. 'That's where I live.'

What a coincidence!

'And we're collecting,' said Maisy, standing beside Joey now, shaking the booty bucket.

A bodyguard stepped forward but Roman thwarted him. 'Collecting for a music school? Now that's a worthy cause,' he said.

Joey's heart thumped. He must keep the conversation going somehow. 'And we've got a fundraising concert. I'm singing the solo. Can you come?'

Honk honk! A driver waved from a stationary limousine. The wiry man beside Roman glanced at his watch. 'We're running late for your next appointment,' he said.

If Roman came, hundreds of people would come too. Maybe Roman could sing a song. It was now or never. Spurred by a shot of adrenaline, Joey said, 'Could you please come, to our concert?'

'Oh yes, yes,' chimed Joey's friends.

Roman shrugged his shoulders and turned to his agent.

The wiry man pulled a calendar out of his blazer pocket. 'What date?'

Sweat stung Joey's eyes. Turtle eggs. He only knew the day. 'The last Friday of this month.'

'Not possible,' the wiry man said brusquely. 'That's Roman's opening night.'

Joey couldn't hide his disappointment.

'Never mind, another time,' said Roman, opening his wallet.

Fifty dollars! That's what he placed in Joey's hand. The note was blue. Joey had never seen one before.

'Enough to buy a violin!' said Ray.

'And here's my name-card,' continued Roman. 'If you ever want a singing lesson, call me.'

The card was bright lucky red. Joey's heart pumped full of joy. 'Thank you. Oh thank you.'

Treeeeeee! The driver had stopped the limousine on the road beside them and a policeman on a motorbike was blowing a whistle, waving it on.

'Okay, okay,' said Roman, as the bodyguards hustled him forwards, past the cordon of adoring fans. Just before he stepped into the car, he turned and winked at Joey. Then the limousine left in a cloud of dust.

'Let me touch it,' said Maisy, pointing to the note in Joey's hand. All his classmates were clamouring, wanting to feel it, and touch the name-card.

'Let's play some more,' enthused Ben.

But Joey was no longer in the mood to sing, and rain drops from a passing cloud were spattering the music.

'We can come again,' said Ray cheerfully, opening his case.

They would come again, Joey thought. On the bus back, he glowed at the memory that Roman had offered him singing lessons. Presumably not for free. He checked the name-card for Roman's address. The street name wasn't familiar but he guessed it was walking distance from school.

And Joey replayed that wink over and over again. It was such a friendly one, so personal. He loved the head nod that had followed it. It was as if Roman was saying: Welcome to the club.

CHAPTER 24

Dreams and Queens

THAT NIGHT, JOEY CALLED his mama to tell her that Mr Waters was helping him complete scholarship application forms. When he'd previously told her about scholarship schemes abroad, she'd reprimanded him for harbouring such foolish hopes.

'He really thinks you're good enough?' said Mama, this time.

'Well, he's going to record me singing and send the tapes tomorrow.'

'Have you eaten?' she replied.

Joey sighed. 'Plenty, Mama. Plenty.'

Why did she always fuss about such trivial matters? Whether he'd eaten, what he was wearing, how often he changed his underwear. He'd be telling her about an exquisite enharmonic change in a Schubert lieder or the joyous cow in *Cow Jumping Over the Moon* and she'd say, 'Put on your jacket, it's cold.' He'd sing an English song he'd composed and she'd ask if he'd finished his homework.

And then Joey realised something. Her world had never touched the Arts. Or English. She'd never performed on stage, or had sleepless nights from rhythms dancing around her head. Maybe she secretly hoped singing was something Joey would grow out of.

It was taking her a while to work out what to say next. 'Let me discuss with your stepfather,' she said eventually.

'Oh, please don't!' Joey cried.

'I have to,' she insisted.

Joey huffed down the receiver.

'Okay,' she said. 'Don't worry about it now. It's late. Wear more clothes. The weather is getting cooler.'

Joey went to bed unsettled. He tossed and turned for what seemed ages, drifting into a restless sleep, which led to a strange dream. He was on a boat with his friends gently sliding down a fast-flowing river in China. Mr Waters, dressed in a flowing white gown, was at the helm navigating the bends of a winding gorge. Monkeys chattered, water lapped, bald mountains pricked a bright blue sky. High on a crag, swallows darted in and out of nests. From its peak, Joey heard a heavenly voice. The unearthly singing floated with the breeze, echoing around, swooping and swirling with the swallows. The song was wordless, a kind of thrumming humming, but it entered Joey's throat and expressed all he'd ever wanted. 'Open your heart,' called Mr Waters and Joey felt his body inflating like a balloon as he joined the joyful melody. More voices entered, even sweeter; weaving, turning, in perfect counterpoint. It was if they were embroidering an ancient silk gown together. Joey felt himself rising.

'Hold him down,' shouted Ray, his ears stuffed with the cotton wool he used when aeroplanes flew over the school. The cotton was multiplying and forming clouds. They morphed into Maisy's and A So's faces, ghostly grey and etched with fear. The sky had turned stormy and his friends were hauling him back. He hit his head on the deck. Then Maisy was tying him down with a rope and Mr Waters was rowing again.

But Joey could still hear the music. It had penetrated the core of his heart, plucking its strings. His whole body was vibrating in harmony and he was rising again.

'He's leaving us, sir,' cried Ray.

Mr Waters rowed faster.

'But he has to sing the solo in the fundraising concert,' said Maisy.

The boat disappeared on the horizon, leaving Joey way behind.

'Wait, wait for me,' he called.

The next day was Sunday. Joey couldn't wait to knock on Miss Asimov's door and give her the busking money hidden under his mattress. He was also dying to tell her about meeting Roman Tam. But the office was closed, and Miss Asimov had gone to the synagogue and would spend the afternoon with friends.

Sundays often dragged and this day was no exception. After lunch, Joey gathered with other boarders to watch a videocassette of Queen Elizabeth II's visit to Hong Kong in 1975. That was the year Mama married the monster. *Hiissss.* The dancing white dots of static disappeared and there was Her Majesty descending steps from a huge plane that had flown all the way from England. A sudden blast of wind blew her patterned dress up. The Queen quickly patted it down. Joey sniggered.

'Hey, that's royalty. Show her some respect,' said Tom, a British sixth-former. 'In England, people stand and sing the British national anthem before a concert begins.'

The monarch was crossing a busy Victoria Harbour on a boat festooned with flags. 'Sorry,' said Joey jumping to his feet and saluting the monarch along with accompanying soldiers. On a boat festooned with flags, she was crossing a busy Victoria Harbour.

Her Majesty alighted at Queen's Pier in Central. Was it named after her, or a Chinese empress? Joey had only heard of a nasty one called Cixi who ordered her mandarins to build a massive boat of marble for her birthday at a time when China was bankrupt and foreign forces were invading. This ruler, Queen Elizabeth II, looked friendly enough, waving to the crowds who were jostling to get a peek at her.

The Queen was now walking down a red carpet towards a makeshift stage. Players in a military band clicked their heels, lifted their instruments and played *God Save Our Gracious Queen, long live our. . . .* that's all the words Joey knew. He *la-la-la'd* to the rest. The makeshift stage was in front of City Hall, where Joey had found the Roman Tam tickets.

Joey glanced at the grandfather clock. Three minutes to four. Miss Asimov still wasn't back. She may not even come into the common room. He'd go to the office first thing the following day.

CHAPTER 25

The Brag

THE DAY STARTED WELL enough. Joey awoke with a new melody in his head. After a breakfast of *congee* and pickles, he went straight to the office. The receptionist told him that Miss Asimov was out until after lunch but she'd let him know he was looking for her.

First lesson: piano. While waiting for Miss Wu to arrive, Joey looked for manuscript paper to notate the music growing in his head but there wasn't any around. Turtle eggs. By the time he found some, the melody's harmonies would probably have vanished into sticky air. There's something I can learn from this, he thought, to cheer himself up. He should carry some manuscript paper in his shirt pocket to write ideas down as soon as they came.

Instead of composing, Joey doodled on the blank page of his piano album. He drew the noteheads, changing them into stick people by adding two legs, four arms, sometimes eight. The semibreves became octopuses and insects.

'Joey, what *are* you doing?' said Miss Wu. Joey hadn't heard her enter.

He quickly turned to his examination pieces.

Miss Wu frowned. 'How are you going to pass if you don't practise?' she said.

'I have practised, Miss,' Joey replied. His gut told him he would pass Grade 5 just fine, probably with distinction.

Miss Wu opened the window to change the air then sat beside him. 'Let's hear your scales then. I'll pretend I am the examiner. That means, I say the name of the scale, or arpeggio and you, *without saying a word,* will play it. If you make a mistake, just keep going. F minor!'

'Harmonic or melodic, Miss?'

Miss Wu tutted.

F minor, harmonic: an annoying, sad little scale which demanded careful control of the distance between the D flat and the E natural.

F minor, B flat major, C sharp melodic minor. How mean could she be?

When Joey played them flawlessly, Miss Wu curled her lips into a smile. 'Not bad,' she said. 'Just remember to keep your mouth shut.'

Joey also played his examination pieces well. As an experiment, Miss Wu took the music away from him for the last piece and he could play it from memory.

'After you've passed Grade 5 Theory, we'll start studying some Grade 8 pieces,' said Miss Wu. 'It'll look good on your scholarship applications.'

'Can you play me one now, Miss?' Joey asked.

Miss Wu retrieved another album from her music case. 'How about *Gollywog's Cake Walk*?'

'I could do with a few cakes,' joked Joey.

He stood behind her to listen to her playing. The golliwog was a jolly black fellow, with a jaunty gait and prone to sudden gaffes. Joey looked forward to learning it.

The shouting from the office could be heard all the way down the corridor. The receptionist was not at her desk. It seemed an inappropriate time to be knocking at the door. Joey peeped through the window. Maestro was sitting at a side table opposite a fat man and an assistant studying a sheaf of papers. 'Two hundred dollars for twenty violins. A hundred for ten violas. Repair of. . . .' the fat man was calling out numbers while his assistant ticked boxes and Maestro nodded. Meanwhile, Miss Asimov was talking to a skinny man wearing a shabby suit. His two sidekicks with crew-cuts sitting on either side of him were nodding their heads. There was also a man wearing an orange shirt with a bold geometric design and bell-bottom trousers. He was standing alone, looking out of the window.

'How do I feed my family if you don't pay your rent on time?' shouted the skinny man. Joey guessed he was the landlord of the building. He felt his cheeks flushing. It seemed the landlord as well as an instrument lender were ganging up against the Asimovs to close his beloved school.

More animated conversation. Joey checked his watch and knocked on the door. Silence. Until Miss Asimov waved him in. Turning back to the angry man and pointing to the stash of crumpled notes on the table, she said, 'That's all we have today.'

The two sidekicks reached forward to start counting.

'Should I come another time?' Joey asked politely.

'No,' said Miss Asimov tartly. 'These gentlemen came uninvited.'

The landlord looked unperturbed. 'If no full pay by Friday, I tell police,' he said to Maestro, in English, so that he could understand.

'But Mr Wong, that's the day of our fundraising concert,' said Maestro sadly.

The landlord took his glasses off, blew on them and cleaned them with a tissue.

'One moment, please,' said Maestro, looking at Joey, who was indeed wanting to speak. The instrument lender's assistant stopped mid-phrase and glared.

'Joey?' Miss Asimov, smiling faintly.

Maestro nodded at her.

Miss Asimov reached for Joey's hand and pulled him beside her. 'Allow me to introduce you to Joey. Joey Kung,' she said, tight-lipped. 'He's a very fine singer whom we found, well, on the street. He's just one of the disadvantaged children who have benefitted enormously from our programme here.'

The instrument lender tapped his fingers and turned his gaze towards the garden.

'A Mozart from San Po Kong,' said Maestro.

'We live in Happy Valley now,' Joey piped.

The instrument lender still looked unimpressed. Clearing his throat, he muttered. 'I can't rent a singer an instrument.'

'Sorry, what was your English name again?' asked Miss
Asimov.

'Cuckoo,' answered the instrument lender.

A strange gurgling sound erupted from Maestro's mouth. 'I
like birds,' he said.

'Mr Cuckoo Chan,' said Miss Asimov sternly. 'My father is
a world-famous conductor. His opinion on gifted children
should be held in high regard. Besides, Joey plays one of your
violins, don't you, Joey?'

'Er, yes,' Joey lied.

The landlord's lackeys finished counting the pile of money.
Unzipping his handbag, Mr Wong stuffed it inside, all the
while studying Joey. There was an awkward silence.

Joey felt emboldened by Maestro's eyes. They looked like
Sparky's when wanting to chase Dong Dong.

'All for you!' Joey said, emptying his pockets. The coins
clinked on the table top like rain on a corrugated iron roof. He
picked up some coppers that rolled on the floor and flattened
Roman Tam's fifty-dollar note. 'Me and my friends raised this
for our school.'

The landlord guffawed.

'We went busking, me and Ray and A So and Maisy,' said
Joey hotly. 'We performed at Diamond Hill.'

And then Joey couldn't stop his tongue. He told them how
Roman Tam had listened, donated that fifty-dollar note, and
he'd seen many more in his wallet. When Joey had told him
about the concert, he'd checked the date with his agent,
and. . . .'

'And?' said Miss Asimov.

All eyes were on Joey now. The landlord's golden teeth
glittered in the sunshine that streamed from the window.

'I sang him a song and he said I sang well and he said he would come to our concert if—'

'Come to the concert, hey?' interrupted Mr Wong.

Joey inhaled sharply. 'Yes,' he replied, his English voice sounding as reedy as an oboe. 'And he said he would like to sing too.'

'Roman Tam sing too?' repeated the landlord.

Joey's ears burned with the heat of the room and the audacity of his lie. Damn it, he'd done it again. He pinched his palm. But Miss Asimov was smiling. Maestro too. They looked so lovely when they smiled.

'That's me in,' said the landlord, slapping his hand on the table. 'Five tickets please. I'll deduct the cost from your outstanding rent.'

'Thank you so much!' said Miss Asimov.

Maestro looked confused.

'Are you quite sure, boy?' sneered Cuckoo Chan.

Joey stood tall but his stomach squirmed as if he'd consumed one too many *char siu* dumplings. 'Very sure,' he said.

There was a clamour of voices.

'Well, the tailor is waiting,' said Maestro to Joey. 'As you're here, let him measure you up for a suit.'

'Yes, sir,' Joey replied.

CHAPTER 26

Music Is in Your Blood

B ACK IN THE DORMITORY, where Ray was practising, Joey
dived onto his bed. His friend lowered his violin from
under his chin. In the rush of air, the sheets of music he'd
clipped to his bedstead had blown to the floor. 'Anything the
matter?'

'Nope,' said Joey, burrowing his face in his pillow.

'Is so,' said Ray.

'Play. Just play,' said Joey.

How could he have told such a porkie pie? He pinched his own cheeks in disgust. The lie had just slipped out. He hadn't planned to tell it beforehand. Should he go and confess to Miss Asimov right now? Would she cry? Would he be sent back to Happy Valley for the beating of his life?

Ray was practising that troublesome passage from Bach's Violin Concerto, as usual. In his head, Joey added the harmony beneath the solo part. The music led him into a confession box, like at the Catholic cathedral where Mr Waters had taken him to an evensong. The confessional smelt woody as red pine. The priest on the other side was Miss Asimov but all he could see was her ear pressed to the metal grate.

Sorry, Miss Asimov. Sorry for lying. I'm so sorry.

Miss Asimov's ear didn't budge but she replied in a major key.

Joey concurred strong, solid. *I'll never lie to you again, I promise.*

Let us pray now, said Miss Asimov. Her voice was so calming that Joey closed his eyes. Her lilting entreaties ebbed and flowed and Joey joined with his own. Their voices rose and fell together, harmonious, consonant. Together they reached an amen.

But Joey had a question. *If there really is a God, He would have heard us, right?*

Miss Asimov told him that God could well be busy someplace else but she would pray again for him, later.

Joey imagined her on her knees, head bowed, her elegant tapering fingers interlocked. *Dear God, Please forgive Joey. He's a good boy really. Amen.*

Ray was playing the final movement now. The music was surer, more strident, like marching soldiers.

I'm never going to lie again, thought Joey. *Never. Never. Never ever.*

But Roman had been interested in the concert. He'd been so friendly. It was that ratty agent's fault that he'd declined. Why did a superstar have to listen to the likes of him? Why hadn't Roman told that skinny little grump to go and jump in the nearest *nullah?*

Yes, that was why Joey had told the lie. He believed that Roman truly did want to come. Cantopop stars were renowned for being late for their performances, and the fundraising concert started at 6.30 pm, much earlier than a pop show. Joey slapped his thigh. With some persuasion, Roman *would* come. Because a star had the power to decide how he spent his time. And if he came, he would give lots of money to the school and the school wouldn't have to close and Joey could carry on studying there and everybody would be happy.

Bach's music was triumphant now. Ray and his violin had overcome all the odds.

And so would Joey. He would go and talk to Roman face-to-face, at his home!

Ray mopped his forehead with a flannel as he listened to Joey's plan.

'I'll go there straight after Maths, and I'll tell the guard I've come for a singing lesson,' Joey explained.

'What if he doesn't let you in?' asked Ray.

Joey had to think about that one. He looked skyward for inspiration. 'I'll bring Billy,' he said.

Ray looked baffled but it was time to run to the next class.

A couple of hours later, Joey and Billy cut class and skipped out of school. Roman's home was indeed only a few minutes away. 'It's number 26,' said Joey. 'Not far now.'

Billy was puffing behind him. 'I can't be long,' he said.

Like all the houses on the street, Roman's was behind a ten-foot-high wall and a guard was stationed in a small booth beside a small pedestrian door within the high metal gate. Joey walked straight up to the Indian and announced himself.

The guard's orange turban wobbled as he shook his head. 'How can I let you in if you don't have an appointment?' he said. There was a red dot in the middle of his forehead, and two more promptly appeared on his cheeks when Billy opened his case.

'It's only a musical instrument,' said Joey.

Billy licked his lips, pursed them and raised his trumpet skyward. *Parp parp,* like a car horn.

'Louder!' said Joey.

PARP PARP!

'Get away from here!' shouted the Indian.

But then Billy was blasting a soldier's marching tune and Joey paced up and down the pavement to its rhythmic beat, swinging his arms from side to side. *Please, please Mr Roman Tam. Can you hear us? Can you see us? Please come outside.*

Creak creak. The metal door opened, from the inside. It wasn't Roman Tam but a young man dressed in slacks and a pink shirt. 'My boss isn't the governor, you know,' he joked, in Cantonese. But he said a few words to the Indian guard who picked up an internal telephone.

BUZZZZ, the metal door opened. 'You're lucky,' said the man.

Roman's garden was even neater and tidier than Sassoon's. A maid was pruning bamboo near a suspended urn, sending a cascade of water into a rock pool where goldfish swam. Rows of potted plants lined the pathway to the house. Joey couldn't suppress a whoop when Roman appeared at the front door.

'You've come for a singing lesson?'

'Sort of,' said Joey. It was now or never. He would sing and then beg Roman to hear him at the concert. He took a deep breath with the first line of *Under Lion Rock* on his lips.

'Hey, wait a minute,' said Roman. 'Follow me.'

The hall was paved in white marble from the ceiling to a parquet floor. A sinister-looking personal guard led them to a sitting room where a maid pointed to a sofa and asked them whether they'd like something to drink.

'My usual,' drawled Roman.

'Coca-Cola,' said Billy.

Roman's drink smelled strongly of ginger. 'For my voice,' he explained, 'and if I were you, I'd save yours. Don't you have a performance tomorrow?'

Joey's mind bounced like a rubber ball. Roman had remembered!

'What's he saying?' asked Billy, drowning among cushions.

Joey stood up. He'd never wanted something so badly. He stared intently into Roman's eyes. 'Yes, I do, and I really want you to be there.'

Roman put a hand on his heart and exhaled.

Joey continued, telling Roman about the landlord and instrument lender. Roman frowned and tutted. 'Of course I'd like to come,' he said, 'but I'm afraid things are not that simple.'

'But aren't you a superstar?' Joey blurted.

Roman laughed. 'Even superstars can't be in two places at the same time.'

'Should I put my trumpet away?' Billy asked.

Joey sat down beside him, feeling hot and bothered. This wasn't going well. A voice inside his head told him to backtrack.

Roman looked deep in thought. He took a few sips from his glass before explaining the importance of pre-performance sound-checks. 'Can you imagine the embarrassment of opening your mouth to sing and discovering your microphone doesn't work?'

Joey shuddered, then tried to hide his disappointment in the exquisite Chinese carpet at his feet.

'Remember the saying, "No money, no talk",' continued Roman. 'It is, alas, very true, especially in Hong Kong. So I'm happy to send a cheque to your school. What's the address again?'

'Oh thank you so much,' said Joey. He was, indeed, grateful but he still wished Roman could come. Even if he wrote a cheque for a thousand dollars, it may not be enough to keep the school afloat. Besides, if people saw him, they'd donate more.

'And never forget,' Roman said, 'your beautiful voice is priceless. School or no school, it may take you far. But you must work hard. If you persevere, you *will* succeed. *Di shui chuan shi,* dripping water penetrates a stone, as my mother would say, haha.' Looking thoughtful, he fingered a golden ring. Roman was famous for his jewellery. He'd wear a different necklace each performance, for luck.

Joey hummed a phrase from *Under Lion Rock*:

> *Together we work hard to create*
> *Our everlasting legend.*

Roman smiled sadly. 'I had to spend years working for a tailor and in a bank before getting a lucky break,' he said. 'For years the only singing I did was in night clubs and bars.'

'With TNT, and Roman and the Four Steps, right?' said Joey, trying to keep the conversation going.

Roman cocked his head. 'You've done your homework, my friend.'

'I've watched *The Romantic Swordsman* many times too, on colour TV,' said Joey.

Roman grinned broadly. 'And what about *Bright Future?* That's the drama which made me famous.'

'You sang the theme song a few years ago, right?' said Joey.

Roman nodded. 'The Year of the Dragon, to be exact. That's when I found my song.'

'And when I found my voice!' quipped Joey, recalling Lion Rock.

The superstar laughed. 'Nineteen seventy-six. Let it go down in history as the year you found your voice and I found my song.'

Billy looked at his watch and nudged Joey, who was breathing fast to keep himself inflated.

Roman stood up. 'Before you go,' he said, 'how about I show you my costumes?

Joey jumped back on his feet. What a rare privilege! He'd never seen anything as beautiful as the gown Roman had worn for the concert at City Hall.

One whole room was a walk-in closet and one of its walls a mirror. Joey glanced at himself in it. He was already almost as tall as Roman, and he'd probably grow taller. He hoped his face would become as handsome as his idol.

Tens of outfits hung on poles in open wardrobes along other walls. Creamy-coloured kimonos, silky pantaloons, tailored Western suits in purple, blue, yellow and pink; a line of matching shoes below. 'I went to Japan last year to study fashion and stagecraft,' said Roman, rippling through his outfits as if they were strings of a *guzheng*. 'This,' he continued, 'is my absolute favourite. As a young man, I was obsessed with Chinese opera. My mama, sister and I would act out the stories for fun.'

Joey gasped. The gown was very similar to the one his papa wore in the photo he had folded in a piece of manuscript paper in his shirt pocket. He hurriedly retrieved it.

Roman studied it carefully. 'How extraordinary,' he said. 'I believe I went to that performance, in Canton. I travelled for two days to reach there and afterwards, slept overnight under a railway bridge. Hey Joey, music is in your blood.'

Something swelled in Joey's heart. God was giving him one last chance. Words tumbled out of his mouth as he told Roman how his mother kept his papa's gown hidden under her mattress, that he would ask his mama if he could give it to him.

A phone rang and another maid entered the room. 'I'll take it in my study,' said Roman. 'Probably my girlfriend,' he said, and winked.

Billy was looking at his watch. 'My driver will be waiting,' he whined.

Sighing, Joey sunk back into the sofa. Billy tapped his trumpet case in irritation.

'I just need to know if he wants it,' said Joey.

'Who'd want some smelly old costume?' said Billy.

Joey threw a pillow at him and Billy reached for one to lob back.

'Hey, Joey,' said Roman, re-appearing at the doorway. 'Thanks, but there's no need to gift me your papa's precious gown,' Roman said, when he came back. 'I'll borrow it, if you don't mind, and get my tailor to make a copy.'

'Of course,' said Joey. 'Can I bring it tomorrow?'

Roman laughed, remaining standing. 'I suppose so.'

'Can we go now?' said Billy.

CHAPTER 27

Flat as a Pipa Duck

I T WAS RAINING. Cats or dogs. Pangolins or pandas. Who cared? Mama said she'd be out working all day but could bring the costume to the concert. Too late! Besides, today was the dress rehearsal, and Joey didn't have time or a key. So Roman Tam wouldn't be coming and Sassoon School of Music would have to close.

Joey glanced at the poster of Roman above his bed and grimaced. He'd like to use one of Mr Clarke's big black marker pens to draw fangs on Roman's handsome face. Give him rabbit ears.

'Wake up, Ray,' he said, shaking his friend's bed. Ray was lying on his back, his violin on his chest. He'd been tossing and turning so much through the night, Joey had worried he'd squashed his instrument.

Ray rubbed his eyes.

There was the splashing of showers, the cleaning of teeth and the billowing of hastily made beds. Joey dressed and went down for breakfast. No one was talking much and he wasn't in the mood anyway. The chit-chat around him dripped down the damp walls like drops of humidity.

After eating, Joey went to the concert hall. He sat on one of the audience chairs, imagining himself performing on the stage. But Miss Asimov was there, checking that the screws of the music stands were tight. Nothing worse than the crash of music stand during a quiet bit in a performance.

'You'll be fine,' she said, putting the folders of music on the stands. Joey knew how carefully she checked through all the string parts to ensure that all the bowings and breath marks were pencilled in properly.

The rehearsal got underway. 'Order!' cried Maestro, tapping his baton on the podium. First up was Bach. Joey hunched into a seat next to Ray's violin case halfway down the aisles and listened. His friend's bow jittered across the strings a few times but the bit he'd practised a million times went well. After then, he looked much more relaxed, playing sweetly in the second movement and defiant as a *kung fu* fighter in the third.

'Was I okay?' Ray was breathing fast. Hugging his violin, he dusted its neck, wrapped it in silk and placed it lovingly inside the velvet casing.

'Probably better than me,' answered Joey. Mr Waters was approaching him. He accompanied him to the stage. 'Pooh, what's that smell?' he said, blocking his nostrils.

Joey wondered what he was talking about.

'Joey, what did you have for breakfast?'

'Chive dumplings, sir.'

Mr Waters covered his nose with a handkerchief. 'Not just before a performance, please!'

'Hah!' Joey exhaled loudly.

'That's not funny,' said Mr Waters.

Maestro was walking back to the podium after writing something in the horn parts. Joey cleared his throat because there was a frog in it. Then he opened his mouth to do a warm-up exercise and only a croak came out. Upturned trees in a typhoon, tower blocks crashing down, cars smashing into shops: without a voice, his path to the future was rubble and wreckage. He imagined people clutching their ears and laughing uncontrollably at the concert. Why on earth did his voice feel as fragile as an eggshell porcelain bowl today? As if, with one crack it could splinter into a million shards. A So had already been banned from singing in the choir because his voice had broken. Had Joey's too?

'Don't be nervous,' said Maestro, patting him on the shoulder.

'Yes, sir,' said Joey. But then, when he sang, from the back of the hall Mr Waters was calling, 'It's not carrying,' and try as he might, Joey couldn't project his voice. He managed to reach the end but there were no cheers or shuffling of feet. Billy blew a raspberry on his trumpet.

Mr Waters came to the stage and Joey overheard him telling Maestro that he'd told Joey to save his voice for the real performance.

Maestro took off his glasses and gave Joey a searching look. 'Here's hoping you pull out all the stops tonight,' he said, tapping his baton against the podium for the next piece.

Joey was furious with himself. He'd sung as flat as a pipa duck hanging in the window of a cooked food stall. He went to the toilets and locked himself inside a cubicle. He rarely cried. He'd wept at his papa's funeral. He'd sobbed as if there was no tomorrow when a sparrow had smashed into his bedroom window and fallen, stunned, onto his balcony. One broken wing was spread like a deck of cards. Joey had cupped it in the palm of his hand for hours, stroked its soft feathers, felt the beat of its little heart. But it had died anyway.

Tears splashed down his nose. He brushed them away angrily. Tears never got you anywhere. He snorted away the rest of the wetness tickling his nose.

But what was he going to do?

Blow his nose. Wipe his face. Go to reception to pick up his suit.

CHAPTER 28

Under Arrest

'I T'LL BE ALRIGHT on the night,' said Miss Asimov as Joey passed her in the corridor.

'Hope so, Miss,' Joey replied, unconvinced.

While waiting for the receptionist to get off the phone, he leant over the counter and glanced at a copy of the *South China Morning Post*. His whole body prickled with adrenaline at the headline:

JEWELLERY HEIST! One Thousand Dollar Reward!

Joey stared at Roman Tam and his heart thumped like a timpani roll as he listened to the receptionist's radio:

We're receiving news that tonight's show may be cancelled. The Cantopop idol refuses to perform until his precious medallion is returned. Roman Tam has a large collection of jewellery but this piece was gifted by his mother and is engraved with a Chinese saying very close to his heart. There is a reward of one thousand dollars for any information leading to an arrest. For the refund of tickets, you can call 5-240975. Let's go now to Roman's agent to tell us more. Good morning Mr. . . .

A medallion. Joey scrunched his eyes trying to remember what Roman had told him about his mother the previous day. What was that saying he'd quoted? Something about water. Joey studied a small photo of the necklace in the *SCMP*. Gasping, he recalled where he'd seen it before – the water character was identical. And that chunky chain was unforgettable. He'd bet his bottom dollar that the saying on the other side was the one Roman had urged him to heed.

But he'd seen it at the pawnshop months ago. Last autumn. Had it been re-sold? Why was it only reported missing now? What should he do? A thousand dollars in his pocket would be fantastic. He'd donate it to the school. But the pop show wouldn't go ahead if the medallion wasn't returned – meaning the superstar would be free to come to the fundraising concert after all. Maybe he should go to Roman's house right now and plead with him. He was expecting his papa's costume anyway!

No. Joey couldn't do that. It was dishonest, calculating and mean. He didn't care about the reward as much as coming clean and being honest. It was already ten minutes to four o'clock. Should Joey call that telephone number? The friendly receptionist would surely let him use the phone.

She did, Joey called, and the line was engaged.

Even better, he'd go directly to the police. There was a station a few minutes run away. He ran fast as the north wind. He was covered in sweat by the time he arrived.

Joey pushed the swinging doors of the police station. A policeman was on duty and Joey told him about the medallion and the man he'd seen selling it.

'But what were you doing at a pawnshop?' asked the policeman. 'Be careful, young man. Anything you say could be used as evidence against you.'

Turtle eggs. Joey felt himself flushing.

'You need to make a statement,' continued the officer, passing him a sheaf of lined paper and a pen.

'I bet it's the one,' Joey said, 'and it had *'Di shui chuan shi'* engraved on the back. That was his mother's favourite saying.'

'That's interesting,' said the officer. 'Sit down over there.' He picked up the phone and dialled a number.

The moony face of a clock *tick tocked* on the opposite wall. Below it, the red glow and burning joss sticks of a wall shrine in honour of Emperor Guan, the god who protects the police, and gangsters. Joey felt more and more uneasy.

The policeman put down the phone, barked down a walkie-talkie and two other policeman entered. 'You're under arrest, boy,' said one of the officers, clinking handcuffs.

'No, no!' cried Joey, as he was led through a door into a room with a row of cells. 'There's some mistake,' he said as he was locked inside one.

The policeman sat himself at a desk and swigged from a flask of tea.

'Hey, Joey!' a familiar voice called.

Joey couldn't believe his eyes. It was Todd, in a nearby cell. Both his arms were tattooed with dragons, his hair was dishevelled and his voice had fully broken.

'What are you doing here?' said Joey.

Todd spoke in a rasping tone, telling him he'd been arrested for a burglary in the neighbourhood, with Sam, the guy in the cell to his left. 'But we didn't do Roman Tam's house,' he said.

'Big deal,' said Joey, feeling his body temperature rising. His cell was dirty and smelt of body odour. 'Let me out!' he cried, to the policeman. 'I'm innocent.'

'Tell that to the judge,' answered the policeman.

'I would, if he were here,' said Joey.

'That you stole the property of another person?'

'But I didn't. I told you. It wasn't me!'

The policeman leant back on his chair. 'Tell you what, young man. All your talking is giving me earache. Could you shut up?'

'I need Roman Tam. I need to talk to him. He'll understand,' Joey replied.

'And I'd like to fly to the moon,' said the policeman, and belched.

There was the sound of footsteps coming towards them, several male voices, the waft of cigarette smoke. The footsteps stopped.

'They're in there,' Joey heard.

'I'd like to identify the medallion first,' said another voice.

'Yes, it's the one,' he heard later.

Joey waited. The minutes ticked away slowly. He couldn't bear the idea of missing the concert. He imagined his classmates polishing their instruments, shaving their cane mouthpieces, cleaning the rosin off their bows. Ray would be practising that passage eight bars before letter B just in case the dress rehearsal had been a fluke. Maisy would be choosing her dress and deciding which hairstyle suited her best. He jumped up and rattled the bars of his cell. 'I must speak with Roman Tam. Mr Roman Tam. Please Mr Tam, is that you? Come in here.'

It wasn't Roman Tam but the pawnbroker. Joey would recognise his straggly eyebrows anywhere. And Roman's agent, holding the medallion! A policeman led them towards the cells and they peered through the bars. 'That's strange,' said the agent. 'This boy was singing outside Golden Harvest a week or so ago.'

The pawnbroker looked bewildered. 'I recognise him too but—'

'What about that one?' asked the policeman.

Todd groaned.

The pawnbroker turned to the policeman. 'Yes, I definitely know him. But I don't remember him selling me anything of much value, apart from a trumpet.'

Joey quickly pieced together what was happening. The pawnbroker must have reported that the medallion was in his shop and contacted the police. Meanwhile, Todd had been caught burgling in the area and was the chief suspect, until Joey turned up and accurately described the wording on the other side of the medallion. Meanwhile, the pawnbroker had been driven over to Kowloon to hand the jewellery over to Roman's agent, who had arrived at the police station at the same time. 'I can explain everything,' Joey cried.

The pawnbroker and the agent were signing documents at a desk only half-listening to Joey. But the policeman was all ears, and was scratching his head. 'But how do you two know each other?' he asked Joey and Todd.

Joey's heart sank. He couldn't get away with it now. But maybe he didn't have to reveal everything. He looked at Todd, who was staring blankly at the wall.

'We went to the same school,' Joey said. 'And I can prove that it wasn't him that stole the medallion. Because even though we often—' He checked himself.

'How?' interrupted the policeman, raising his eyebrows at the agent.

Joey repeated that he was certain that the man who sold the medallion wasn't Chinese.

'Well, my memory is not as good as it used to be,' said the pawnbroker, leafing through a ledger he'd retrieved from his carry bag. 'But this should help.'

The agent nodded. 'The boy's account does make sense,' he said. 'Roman keeps all his jewellery in a safe and only I have the key. He hasn't worn his mother's medallion for as long as I can remember but he always wears the matching ring. And one of his staff was dismissed last autumn for dishonesty.'

Joey sighed with relief. Surely it didn't matter who the original thief was. The significant fact was that the medallion had been found. Did that mean that Roman's show would still be cancelled?

'No, it will definitely go ahead,' said the agent. 'Preparations have started and the supporting band can still play. But Roman may sing a little later than originally scheduled.'

Joey's mind was a hotbed of ants. 'So please, sir. Help me. Does that mean Roman would have time to come to my school's fundraising concert and hear me sing?'

The agent looked taken aback. 'Well, I suppose. . . . Let me phone. . . .'

'You're a singer too? I don't believe it,' said the policeman.

Joey cleared his throat, ran his tongue along his teeth to coat them with saliva and licked his parched lips. Taking a deep breath, he began to sing:

> *Life has its challenges*
> *It's not without its worries*
> *In the same boat under the Lion's Rock we row together*
> *Putting aside our differences and finding common ground.*

The policeman slapped his desk. 'Love it!' he said. Joey sang louder. He opened his chest and filled his lungs to the full:

> *Putting aside our hearts' conflicts*
> *Together we pursue our dreams*
> *In the same boat we promise to go together*
> *Without doubt or fear.*

The policeman was singing along with him. The pawnbroker too, as well as the agent when he re-entered the room. Todd stayed hunched on the bench of his cell while Joey pranced around pretending to be Roman Tam. Everyone clapped at the end.

'You've got a good voice, young man,' said the policeman, unlocking his cell.

'And I've got some very good news for you,' said the agent. 'Roman is delighted that it was you who found his medallion, and says he's willing to come to your concert.'

'Hurrah!' shouted Joey, punching the air. He gave Todd a backward glance before racing back to school.

CHAPTER 29

Pre-Performance Antics

A LOW-FLYING AEROPLANE droned overhead as Joey wended his way between pedestrians and traffic. With luck he'd be back in time to have a snack. Flashes of joy shot up and down his spine at the thought of sharing his great news with Miss Asimov.

Hereford Road was congested. Taxis and private cars slowed at the kerb to let people alight. A foreign lady with permed hair peered at him from inside a car with government number plates. A Chinese man with a moustache held a door open as an elegant stockinged leg appeared from it. Of course! They were coming to the concert. Roman Tam would come too. Joey's stomach squirmed.

But hey? What was happening at the school gates? They were wide open and a row of ladies, mainly Westerners, were strung hand-in-hand across it facing the road, where Miss Asimov and the receptionist were standing. Joey slowed and stopped to listen. They were talking about the possible imminent arrival of the instrument lender.

'Goodness me!' said a lady in a straw hat with ribbons.

'Ridiculous!' said another.

'How inconsiderate!'

'Scandalous!'

'Report him to the *SCMP!*'

A well-rounded lady with rosy cheeks in the middle of the group passed her handbag to her much shorter husband, saying, 'I've got a few tricks to counter the likes of Mr Cuckoo Chan. I didn't work behind the lines in the War for nothing.'

'Hear hear,' said a woman in stilettos. 'Terence, look after my umbrella would you, darling?'

'Keep it. It's rather overcast,' her husband replied.

'It'll serve as a good poker too,' said the red-cheeked lady. 'Never fear, Nurse Perkins is here, ready to mop up some blood.'

'Save a seat for me in the hall, honey,' called the woman in stilettos as her husband walked down the garden.

Joey tapped Miss Asimov on the shoulder. She swung round, and gasped. 'Joey! Where *have* you been? Go inside immediately.'

At that moment, a Chinese couple came forward to greet her and she extended a hand to welcome them.

'But Miss Asimov,' said Joey, 'Roman Tam will come to the concert. The Cantonese pop star . . . please listen to me. . . .' His voice faltered as more guests gathered around her, commanding her attention. Joey wasn't sure she'd heard him.

'Miss Asimov,' purred a Chinese lady, 'I hope you don't mind me saying but I think you've lost some weight.'

'I have?'

'Running a charity takes its toll, doesn't it?' added an elderly man.

'Where's the collection box?' called a fine lady, waving a large manila envelope.

'Mine's here too,' said a man, delving in the inside of his suit pocket.

Just then, the landlord, his wife and three heavily decorated girls alighted from a pink Rolls-Royce.

Miss Asimov raised her arm. 'Ladies and gentlemen, thank you all so much. My father and I truly appreciate your kindness. Donations will be collected inside but . . . but . . . oh no!'

The barricade ladies were crying and pointing and Joey turned towards where they were looking, towards a slowing truck, a truck with CUCKOO CHAN MUSIC SHOP painted in large letters on its side. The squeal of its airbrakes made the ladies jump.

'It's him,' said Miss Asimov despondently.

'I'll call the police,' said the receptionist.

A worker jumped from the truck's passenger door, walked to the back end and banged it rhythmically with a stick while the driver, cigarette dangling from his lips, revved the accelerator, cursed and reversed into a narrow parking space.

Cuckoo Chan appeared from the cab and walked towards the gate.

'Miss Asimov, let me handle this,' said Nurse Perkins firmly, and the barricade of ladies gripped their hands tightly. Joey joined the end of the line.

Cuckoo Chan tried to pass. *'M'goy,'* he said brusquely.

The landlord and his family were behind him, trying to enter too. 'Mr Chan, what's going on here?'

'No!' shouted Nurse Perkins.

Cuckoo Chan smiled coyly at the ladies, most taller than him. *'M'GOY?'* he said sweetly. His workers were now grouped around him. Strong skinny men with tattoos on their bare backs and chewing toothpicks. *'Maa gwai faan,'* said one of them, and spat.

'Disgusting,' said the lady in stilettos.

Cuckoo Chan straightened himself and raised his chin. Smiling a fake smile, he began, 'We here to. . . .'

'NO!' Joey joined a chorus of voices.

Cuckoo Chan broke into ugly Cantonese, a language he knew Miss Asimov understood. 'Get these *gwai po* away from here. You want trouble? I give you trouble. I call more men. I call police. Yes, I call police.'

'Call police,' repeated one of his minions.

Miss Asimov pursed her lips. 'They've already been called,' she said.

Stalemate. Everyone paused for breath.

Then there was the roar and flashing of police motorcycles. 'Move aside,' called one of the officers, brandishing a baton.

'What they doing illegal,' shouted Cuckoo Chan.

'But he's come to take instruments away, just before our concert,' cried Miss Asimov.

The landlord and his daughters had joined the scrum. 'She owe me thousands dollar too. Two month!' he shouted.

There was some jostling to break the line but Joey and the ladies maintained their positions.

'We have our orders!' shouted a policeman.

'Help!' cried Joey, as the knees of the lady beside him buckled and she dropped to the ground.

'Coming,' called Nurse Perkins.

'Ladies and gentlemen, PLEASE!' cried Miss Asimov.

Her words went unheeded in the flurry of pushing and shoving.

'Joey, Joey. Come here. Listen to me,' she called.

'Yes, Miss,' said Joey, sweat pouring down his face.

'Go and tell Maestro what's happening.'

Joey bolted inside.

CHAPTER 30

Airborne

THE CONCERT HALL was decorated as if for a birthday party. Coloured balloons were strung on the awnings, bopping against the whirring ceiling fans. Most of the seats were filled. There were queues for the bathroom, programmes and interval drinks. But no sign of Maestro.

Near Joey, at the back of the hall, mums unpacked Tupperware and unwrapped silver foil. Scones, ginger biscuits, cheese sticks and cherry pies from Western mums. Mango puddings, sago, crab sticks, seaweed and spam from the Chinese. Students were dipping their fingers into a trifle when Yum Tai, the tea lady, wasn't looking.

Ding! The three-minute-to-performance bell. Joey slipped behind the velvet curtain, quickly changed into his suit and waited stage left. His classmates were backstage too, lined up to enter, laughing, joking, tuning, blowing. Mary the oboist was making a last-minute reed change. Ray was practising that passage which still bothered him. The sight of all his school-mates thrilled Joey. But where was Maestro? Everyone he asked hadn't seen him.

Ding! The one-minute bell. All orchestral members were on stage now and A So waved at Joey. Then, in that long-established tradition, the strings, woodwind and brass merged into a verdant A. Gut strings, drum skins, ebony wood and metal, all vibrating in perfect harmony. Soon now, the lights would be dimmed, the audience would hush and a rush of goodwill would warm the air.

The stage hands were poised to open the curtain after Maestro's introductory speech. Suddenly he appeared at the opposite side of the stage, stroking his baton. Joey raced over, jumbling his words while trying to tell him about what was happening on the street, and Roman Tam. Maestro patted his head. 'No time, dear boy,' he said. 'What will be will be. It's in the lap of the gods.' He walked in front of the closed curtain and the audience clapped.

With a beating heart, Joey listened from the wings.

'Ladies and gentleman,' Maestro began. 'In a few moments the performance will commence. As many of you know, this concert is a special one. Not only are we showcasing our most promising students but also, in the light of—'

Clomp. Clomp. Joey covered his face with his hands. The workmen's boots were stomping up the back stairs.

'. . . Sassoon School of Music first opened its doors in 1970, the year I—'

Crash! Bang! Cuckoo Chan's men barged on stage pushing trolleys, Miss Asimov hot on their tail. One of them started taking percussion instruments. 'Go away!' cried Miss Asimov.

Maestro droned on, oblivious. 'This hall has a fascinating history. In early colonial times it served as a customs office—'

CRASH! BANG! Maestro paused and cocked his ear. At that moment, one workman was grabbing a violin from a small girl at the back of the violin section and another was pinching a clarinet.

'Maestro,' called Joey, but the conductor continued, '. . . and so your generous donations would be much—'

'Get off, it's mine,' screamed a viola player.

'Go away, it's hers!' shouted Ben.

'Pack your cases and run,' shouted A So.

Miss Asimov was having a tug-of-war with a man and Maisy with her cello.

Joey had to do something. He pushed the curtain aside and walked towards Maestro, blinking in the strong light.

'Excuse me,' Maestro said to the audience, looking anxiously at Joey.

But then, above the cries and shrieks and pulling and shoving of instrument-collecting behind the curtain, there was a collective gasp from the audience.

The removal men froze, listening to the whistles and cheers and calls for a collective shush.

'Thank you, thank you,' said a fruity voice.

Was it? Joey's heart skipped a beat.

It couldn't be. Could it?

It was!

Roman Tam. Walking up the aisle, bowing to the audience, smartly dressed in a Western suit and purple-striped cravat.

Miss Asimov was running towards Joey, breathless, beaming.

'You did it! You did it!' she whooped, wrapping him in her arms. It felt so good.

'For our school,' said Joey triumphantly.

When Roman walked on stage, the removal men scattered.

'Roman, Roman,' the audience were chanting.

'How about *Under Lion Rock?*' he said, loosening his cravat.

The curtain whirred and jerked as the stage hands opened it. Maestro stepped on the podium and the orchestral players hurriedly found their parts. Then the honeyed tones of Roman Tam were floating through the air.

> *Life has its challenges*
> *It's not without its worries*
> *In the same boat under the Lion's Rock we row together*
> *Putting aside our differences and finding common ground*
> *Together to the ends of the earth.*

'Bravo!'

'Encore!'

'Encore!' cried Joey.

And the hall erupted into cries for more.

'Thank you, thank you!' called Roman Tam. He and Maestro took three curtain calls. Joey hadn't noticed Maestro's slight limp before. Third time, baton held high, the conductor pointed to Maisy, the marvellous cellist, and the audience cheered. He shook hands with Ray, the leader of the violins, and Ben, principal viola. Then he gestured towards the wood-wind and brass. Billy grinned.

It seemed Cuckoo Chan wasn't happy to be defeated. The moment the clapping subsided, he sidled up to Roman. But the superstar shrugged and turned his back on him. 'Donate as much as you can,' he called to the audience, walking down the aisle towards the exit. 'You will not find a worthier cause.'

Maestro was calling Joey. 'Your turn,' he said.

'What now?'

'I think that would be best,' replied Maestro.

'Fine by me,' called Ray.

Joey moved centre stage next to the podium and the orchestral players whooped. Maestro tapped the conductor's stand for attention and everybody hushed. You could have heard a feather drop.

The lights felt warm on Joey's back. He looked behind him.

Miss Asimov, waving and smiling from stage right.

Maisy, blushing.

He turned his head and looked out at the sea of faces. His throat tickled and his mouth felt as dry as a preserved plum. Could he sing? Would his voice crack? Panicking, he searched in the darkness for the one he needed most.

There she was. His mama. Their eyes locked, or so he thought, and she waved shyly. Sitting stiff beside her, Step-father, was studying the programme. And there, in the row behind them, was Aunty Tam, Uncle Bo, Aunty Ma, the bean-curd man, the beef man, and Sparky! The little dog was sitting on his knee. And on the front row, a stone's throw away, was Mr Waters, Miss Wu, and all his other teachers, smiling.

What about Roman Tam? Was he still there?

Yes! He was signing autographs at the back of the hall.

What did Joey have to be scared of? He squared his shoulders and cleared his throat.

Maestro lowered his arms to give a downbeat. A golden harp strummed. Mary's sweet oboe entreated. Maisy's cello thrummed, calming Joey's beating heart. He envisioned hawks hovering high above Victoria Harbour, his voice stretching its wings far and wide, higher and higher.

In a flash, Joey was sure of what he hoped for and certain of what he could not yet see. His lungs were clear and victory coursed around his body like blood.

In the corner of his eyes, he saw Maestro's cue.

He took a deep breath and flew to where the angels sing.

HISTORICAL NOTE

In the 1970s, Hong Kong experienced rapid societal, industrial and economic change. As the city developed into a major international entrepôt, people from all over the world came to seek their fortune. Hundreds of thousands of immigrants from Mainland China and Southeast Asia arrived too, adding to the cultural mix.

Hong Kong had benefited from being a diverse and cosmopolitan city since the mid-nineteenth century. Among the many immigrant groups were White Russian Jews, like Maestro Asimov and his daughter, and Iraqi Jews, like the Sassoon family.

The Sassoon brothers existed and their legacy continues to this day due to their generous donations to hospitals, schools and other charitable causes. But 'The Sassoon School of Music' is a figment of my imagination. So too, are Maestro Asimov and his daughter. *Under Lion Rock* is thus 'historical fiction', a literary genre that blends fictional elements with real-life historical events, figures and settings.

Real characters in this story include the superstar Roman Tam (1950–2002), who is generally recognised as the 'Godfather' of Cantopop. Queen Elizabeth II, Lim Kek-tjiang, Sam Hui and Yo-Yo Ma are genuine too, although only Sam Hui and Yo-Yo Ma are still alive. The other characters are fictional.

Lion Rock is a mountain in Hong Kong named for its distinctive shape that resembles a lion. After World War II immigrants built makeshift homes, 'squatter villages', on mountainsides due to the lack of affordable housing in the territory. The Hong Kong TV series *Below the Lion Rock* began in 1972 and told the life stories of different social groups set against backgrounds that are today part of Hong Kong history. The theme song *Under Lion Rock* (1979) (or *Below the Lion Rock*) became one of the most famous Cantopop songs ever written, and recalls the 'Lion Rock spirit', a term that arouses pride for the perseverance and resilience of Hong Kong people. For example, the song was broadcast extensively during the outbreak of SARS in 2003, as well as in 2013 for the government's 'Hong Kong Our Home' campaign.

Some of the settings of my story still exist, others don't. The squatter village on Lion Rock, the Golden Harvest Movie Studio, Macau ferry night market, Queen's Pier, Blake Pier, and Lai Chi Kok Amusement Park have long gone, while the Wholesale Fruit Market, City Hall Concert Hall, Happy Valley Racecourse, Hong Kong Star Ferry, the Bird Market and Hereford Road are still here. So is the Aberdeen Typhoon Shelter, although most of the thousands of 'boat people' who lived there have since been resettled.

Jane Houng
June, 2023